Violin Acoustics

A New Look at the Old Art

Augustus F. Matthews

VANTAGE PRESS
New York / Los Angeles / Chicago

To violin makers
and everyone interested in the truth
about violin acoustics

FIRST EDITION

Published by Vantage Press, Inc.
516 West 34th Street, New York, New York 10001

Manufactured in the United States of America
ISBN: 0-533-07594-7

Library of Congress Catalog Card No.: 87-90206

Contents

CHAPTER 1

Familiarization and Introduction

It is believed that the old Italian violin makers had some method of tuning their instruments and therefore had a knowledge of acoustics as applied to the violin not fully known today.

The writer has spent a number of years researching this tuning process and acoustical principles as obtained by the old Italians in their work of producing exceptionally resonant instruments. It is the writer's conviction that this research and experimental work has produced knowledge of the tuning and acoustical principle that is similar or possibly the same as that known and practiced by the old Italians.

No attempt has been made to give instruction on the construction of violins. There are excellent books on violin making available at bookshops.

The work here is exclusively on tuning and acoustics. It is intended as a companion work to the violin making books now available.

We have many fine craftsmen today whose work is quite the equal in craftsmanship of the old Italians. A knowledge of the tuning process and the true acoustical principles involved as known by the old Italians would make our modern makers quite the equal of the old Italians or even better than they were.

Excellent construction is accomplished by our modern violin makers by following the patterns and construction lines, graduation thicknesses, and all other construction processes enumerated in violin instructions and by following the construction lines of the old Italian instruments. Stradivari and other old Italian violins have been completely examined and the construction lines fully written about in books available. All our modern violin makers need to know is the true acoustical principles for maximum resonance and tone quality.

Those who wish to verify the step-by-step principles as stated here may do so, as every experimental finding proves itself. The experiments as conducted by the writer may be done by anyone.

The writer wishes to state that all the experiments and tuning were done with a set of twenty-one tuning forks. A set of thirteen forks, as those used for tuning pianos, in one-half tones from C=261.63 to C=523.3. Also, several additional forks were especially made from G=196, G# =207.65, A=220, A# =233.08, and B=246.91.

Tuning forks are a wonderful tool for the tuning process. Every violin maker would find great help in using the thirteen forks as used for tuning pianos as well as A=220, A# =233.08 and B=246.94.

There is now available an electronic tuner, which is much cheaper, possibly half the price of tuning forks, and is used as much, or more than tuning forks. Tuning forks and electronic tuners may be purchased from Turner Supply Co., 944 Wheatland Ave., Somerville, Massachusetts 02143. The phone number is (617)-666-4550.

The electronic tuner is quite adequate for tuning the violin plates. The tuning forks have some advantage in that they can be sounded directly on the plates for determining the intensity of resonance. However, this is not necessary for simple tuning.

The violin is tuned to the lowest octave, from G=196 to G=392. It is accomplished by tuning the several parts of the violin to specific frequencies. The back plate, the top plate, the air column, the bar, and even the sides are used for the lowest frequencies.

The back plate is tuned to A=220.
The top plate is tuned to B=246.94.
The air column is tuned to C# =277.18.
The bar is tuned to F=349.23 or F# =370.
The sides give additional area for the lowest tones.

The frequencies of the top and bottom plates by percussion change somewhat when the final graduation is done. This will be discussed later when the graduation is made. The tuning of A=220 for the back plate and B=246.94 for the top plate is based on the plates of equal thickness of 8/64 of an inch for the top plate and 10/64 of an inch for the back plate for testing by percussion to determine the frequency.

The selection of the wood is most important, and a method of testing the wood for its resonant qualities must be made. All of these tuning processes are discussed as we proceed with the acoustics of the violin. If any of the parts of the violin are not tuned specifically or the selection of the wood is not adequately made, the quality of the tone will suffer by that much.

CHAPTER 2

Selection of Wood

The first consideration is the selection of wood. All violin makers know, of course, that spruce for the top plate and maple for the back plate produce the best resonance. However, there are differences in the resonating qualities of these woods. It would be so very helpful if some test could be made on the wood before construction begins.

At this point, reference is made to an old work by J. F. Fetis, published in 1864, titled *Notice of Antonio Stradivari*. In this work, he quotes from the experimental work done by Felix Savart, who worked with M. Villaume on a number of experiments on the violin. He makes some very startling statements, which are hard to believe. Among other observations quoted in his book is an experiment to determine the frequencies of maple and spruce rods taken from Stradivari instruments, dated 1690, 1717, 1724, and 1730. These rods were tested for their frequency by putting them into vibration with a bow. It was an experiment that would be quite impossible today, and it is most difficult to believe it was done then. Who would destroy four Stradivari instruments to obtain the frequencies of these small rods? Any Stradivari instrument, no matter how damaged, would be repaired today.

These fragments were cut parallel to the grain. The dimensions are as follows:

20 centimeters × 20 millimeters × 5 millimeters.
The maple gave a frequency of A# = 456.12.
The spruce rod gave a frequency of F = 683.08.
Seven half-tones between A# and F.

They used the old pitch of 512 = middle C, which would place the frequency slightly below our pitch of C = 523. It is not enough, however, to be very noticeable by percussion.

They do not say from what instruments they took these rods.

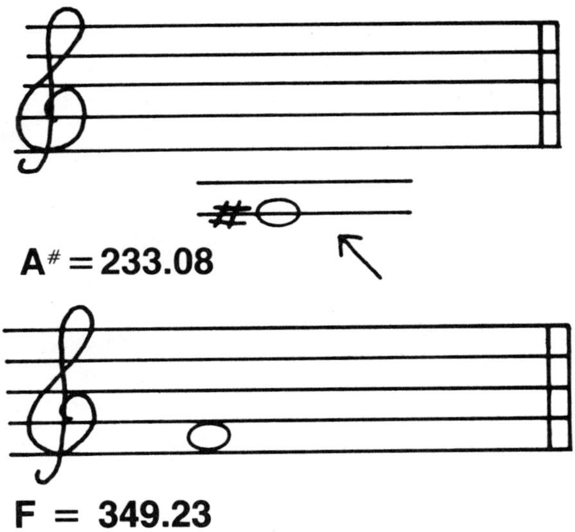

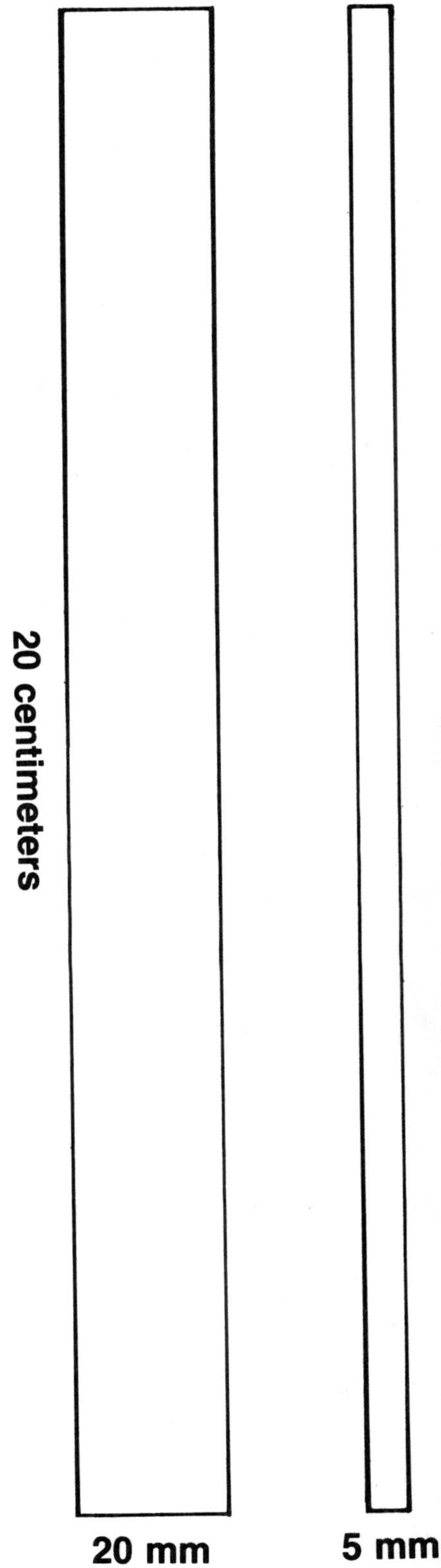

87 *The tone and frequency of rods 20 centi-*
87 *meters by 20 millimeters by 5 millimeters, by*
87 *percussion.*

There are no areas in a violin from which these dimensions could have been taken. It was possibly a cello or a double bass.

They determined the frequency by bowing them, which was not necessary, as it is quite simple to test them by percussion. There is a node about one-quarter of the way down from the end of the rod that can be held between the finger and thumb, and by holding the rod up to the ear and gently striking it with the right hand fingers at the center or near the edge of the rod, the tone is quite discernable. Check with the tuning forks A# and F.

This testing of rods by percussion is explained in the acoustical work by Lord Rayleigh and Chas. A. Culver, and they also give illustrations. The exact location of the node is 0.2204 inches in length.

Savart obtained the incorrect tone. They obtained A# = 456.12 and F = 683 on the pitch of middle C = 512. The correct frequency by percussion is an octave lower at A# = 228.06 and F = 341.72 on the pitch of C = 512.

In cutting the rods, care must be taken to have the thickness exactly 5 millimeters. Indicating calipers should be used. The fact that the same frequencies were obtained from such widely separated times of construction must mean something.

The writer has made a number of these spruce and maple rods. Many of them came out at A# = 233.08 for the maple and F = 394.23 for spruce on our own pitch of A = 440. However, some of them gave higher or lower frequencies, in which case slightly thicker or thinner plates would be indicated in order to obtain the fundamental frequencies of the plates, which we shall discuss later. It is best to use only plates that gave tones of A# and F.

In an effort to find out why there should be a difference in the tone and frequencies of the same woods and why the rods should give different tones, an expert in the marketing of different grades of pine, taken from the slopes of the Sierra Nevada mountains in California, was consulted. The writer felt foolish consulting a lumber merchant about wood for violins, but it was decided to "leave no stone unturned" to find out the truth. The information obtained from this lumber specialist was most revealing, and we must be forever grateful to him. Pine is very similar to spruce in resonating qualities. In fact, this is the wood the older writers refer to as *deal*. After we explained our problem, he thought for a moment and then remarked that he might be able to give some help. He explained that there were three different grades of wood that grow on the slopes of the Sierra Nevada mountains. That which grows on the lowest elevations and receives a minimum of moisture and snow runoff grows very slowly, and as a consequence of this, the wood is hard. The trees that grow on the highest elevations receive too much water and snow runoff. In this case, the wood grows too fast and this wood is very soft. However, there is an intermediate location where the trees receive normal moisture conditions and have normal growth. *This wood is the best grade.*

"Now," he said, "I can give you a plank from all three of these grades of wood; possibly from them you can experiment and find out something." The writer proceeded to cut rods from the three planks of the dimensions as mentioned, 20 centimeters × 20 millimeters × 5 millimeters. The experiment was most revealing. The sample from the hard wood that had received a minimum of moisture was several half-tones higher than F=349.23 at G# =415.3, and the tone was thin and tinny. The rod cut from the plank that was grown in the higher elevation where there was too much moisture and snow runoff was several half-tones lower at D=293.66, and the tone was so dull as to be almost indistinguishable. However, the rod cut from the plank where the moisture conditions were adequate for *normal* growth was exactly F=349.23, and the tone had a clear bell-like ring, by percussion. Similar tests were made on maple rods with the same results. Some were A# ≐233.08 and others were lower and higher in frequencies, which indicate that the same growth conditions apply to maple.

It is clear from these tests that both spruce and maple must be grown under the favorable conditions as outlined here to give maximum resonance.

It is obvious that the old Italians were aware of the necessity for obtaining their wood from trees that were grown under normal growing conditions, and they were willing to travel a great distance to obtain it. It is believed that they traveled to the Italian Alps near the Swiss border for their wood. The spruce certainly could have been obtained there, and possible also the maple. This would mean a trip from Cremona of some fifty or sixty miles.

There is another growth condition that could possibly have desirable effects upon the wood. It came to the writer's attention that wood selected for the purpose of making ladders for fire fighting, namely, Douglas fir, was selected from trees grown in locations where they obtained adequate sunshine, which apparently made the wood stronger. It is quite possible that wood for violins must be selected from trees that received not only normal water conditions, but also adequate sunshine.

Having found that these rods can be a test for the conditions of the wood for violin making, it would be advisable that this test be done before proceeding with finishing the plates. When drawing the outline for the plate upon the spruce top and maple back, it will be observed that there is an area at the upper bout where a rod of the dimensions stated can be cut, 20 centimeters × 20 millimeters × 5 millimeters. It should be determined that the tone is $A^{\#}=233.08$ for the maple rod and $F=349.23$ for the spruce rod. The tones could be slightly lower on our pitch of $A=440$. The thickness must be exactly 5mm. Indicating calipers should be used to be sure of the thickness.

Modern violin makers have a problem in the procurement of their wood. We must purchase our wood from dealers who import it from Europe, mostly from West Germany. We do not know, and they do not know, from where it was obtained. It could be from wood grown under any conditions. Therefore, it is of the greatest importance that the wood be tested by cutting these twenty-centimeter rods to determine the frequency and quality of the wood.

We are told that the old Italians traveled long distances to obtain favorable wood. They may very well have cut down a whole tree, having found one growing under favorable conditions, in which case each craftsman would have enough wood to make a hundred violins or more.

The writer is convinced that any secret they had to produce exceptional tone quality was as much in the acquisition of favorable wood, together with their knowledge of the scientific tuning principle that produced exceptional resonance and tone quality.

The writer has determined that in spite of all the tuning and scientific application for resonance, unless we have wood grown under the favorable conditions as outlined here, maximum resonance will not be obtained.

Before going into the thicknesses and graduating of the plates, it is necessary to find the square inch area of the plates. The directions for this follows.

CHAPTER 3

Determining the Square Inch Area of Plates

For tuning and calculating the fundamental frequency of the top and back plates, it is necessary that we determine the square inch area. This may be done by drawing the outline of the plate on graph paper ruled in ⅛ inch squares. The ⅛ inch squares may then be counted and transferred into square inches.

It will be found that there are 5574 ⅛ inch squares in each plate, which transfer into 87 square inches. When counting the squares, thirty additional squares have been added to the count on each side of the plate at the center bout to include the area for the arching.

It was determined that calculating the area of the plates to the outside edge is best. It was found upon experiment that the vibrations for the lower frequencies stretch out for additional area into the sides; therefore we may calculate the area of the entire plate, which is best for testing the fundamental tone of the plates.

It was thought beneficial to determine the square inch areas of the upper and lower bouts, above the bridge and below the bridge. It was found that there is an equal area above the bridge line as below the bridge. We find 43.5 square inches on each side of the bridge. However, upon closer observation, it was found that the actual line of equal area is on a line slightly below the sound post line on violins with the upper bout containing a width of 6 ½ inches. This has considerable significance. In testing with the tuning forks, we find that the strongest resonance response is on this line where equal area is obtained. Therefore, when the equal area line is placed on a line slightly below the post, the resonance is slightly softened. It is in no way deteriorated, but is merely softened slightly. Therefore, for the strongest resonance, the line of equal area should be on a line between the bridge and post line. This would necessitate increasing the upper bout from 6 ½ inches to 6 ¾ inches.

May we believe that because in the old days most music was for chamber and church music, the old Italian makers purposely placed the equal area below the bridge line to give a slightly softer resonance for this purpose? Certainly increasing the upper bout width to 6 ¾ inches gives the maximum strength to the resonance. We find that Stradivari did indeed make some of his violins with a 6 ¾ inch upper bout.

The writer refers to the line where the strongest resonance is obtained as "the focal point of resonance."

CHAPTER 4

Rectangular Plates

Before going into the thicknesses and graduation of the plates, it was thought beneficial to start by making rectangular plates containing the same area as violin plates. It was considered that valuable information could be learned from rectangular plates. The area of the plates was made to contain the same area as a violin, 87 square inches. The dimensions of the plates are 14 × 6.22 inches—87 square inches. This is the overall area of the violin plates, including the area of the end blocks, the corners, and overlapping edges. The width of 6.22 inches is about the average of the three widths of the plates.

The thinner the thickness of the plates, the lower the frequency. The thicker the plates, the higher the frequency. Therefore, it should be possible to find a divisor for each half-tone, which when divided into the frequency would give the thickness for each half-tone. This has been done, and a list of half-tone frequencies in 1/64 inches and also in millimeters is given for our pitch of A = 440. There is also included a list of thicknesses in 1/64 inches for the old pitch of A = 430 which is said to be the pitch used by the old Italians.

In order to start experimenting with these rectangular plates, let us take the frequency of B = 246.94, which is the fundamental tone of the top plate, with a thickness of 8 1/64 inches. We find by experiment that spruce, rectangular, 14 × 6.22 inches, 87 square inches, and 8/64 of an inch thick gives a fundamental tone of B = 246.94. The divisor is 30.85.

The back plate is tuned to A = 220. The divisor is 22 into A = 220—10/64 of an inch. For millimeters, divide the thickness of 10/64 of an inch by 2.52—3.96 millimeters.

The fundamental tone of these rectangular plates may be determined in the same manner as we tested the twenty-centimeter rods. There is a node about one-quarter of the length from end of the plate. Hold between the thumb and finger and gently strike the plate at center or off-center toward the side. The tone is very clear. Should the tone be not very clear, move the plate up or down slightly until a clear tone is obtained. Verify with the tuning fork.

In testing the plates and rods for tone and frequency, the temperature should be 70 to 72 degrees Fahrenheit. The importance of temperature cannot be expressed greatly enough. Lower temperatures can raise the tone by one-half tone. When you go into a cool room and raise the temperature to 72 degrees, even if the plates still feel cool, the tone can be one-half tone higher.

The importance of temperature cannot be stressed too greatly. For instance, if one leaves plates in the workshop or room overnight and then upon entering the room in the morning and heats the room to seventy to seventy-two degrees, even though the room is warmed, the plates to be tested for tone may remain cool to the hand. In this case, the cool plate will be one-half tone higher than true frequency. Be sure the plate is not cool to the hand. The spruce top plate graduated to 8/64 of an inch with true frequency of B=246.94 will be C=261.63 if the plate is cool. Likewise, the maple back plate graduated to 10/64 of an inch with true frequency of A=220, when cool will have a frequency of A#=233.08.

Tuning fork manufacturers guarantee accuracy at seventy to seventy-two degrees.

CHAPTER 5

Graduation of the Plates, Violin Shaped

Having discussed the experiments with the rectangular plates, we may proceed with the graduated thickness necessary for the vibrational process for the range of all the frequencies. Refer to the lists of thicknesses given in 1/64 of an inch for each half-tone on plates of equal thickness. It will be noted that in order to tune the back plate to A = 220, the thickness is 10/64 of an inch. When the back plate is graduated from 10/64 of an inch to 6/64 of an inch at the perimeter, a paradox is noted. The center goes up to C = 261.63 and the lower ends go down to G = 207.65. However, when the plate is turned the other way and held by the small end, a strong tone of A = 220 can be observed.

It would be advisable to graduate both plates to these thicknesses all over to start with in order to obtain the fundamental tone of A = 220 and B = 246.94 by percussion.

At this point, when the final graduation is done, it must be explained that something of a paradox presents itself. We should think that when the final graduating is done that the tone and frequency would be lowered, but this is not the case. When the top spruce plate is graduated from 8 1/64 inches down to 7 1/64 inches at the perimeter, the fundamental tone at the center, by percussion, will remain B = 246.94. However, there is a much lower frequency at the lower ends of the plate that goes down to F = 174.61. With the back plate, the paradox is even greater. When the plate is graduated from A = 220, 10 1/64 inches down to 6 1/64 inches at the perimeter, a considerable change takes place. The center frequency, by percussion, goes up to C = 261.63, and the lower ends go down to G# = 207.65. This testing is done on the plates before they are glued to the sides.

The writer feels sure that the truth of these frequency changes will be questioned. It is almost unbelievable, but it has been tested over and over again, and no other conclusion can be reached.

There is another tuning response that should be mentioned, with regard to the back plate. In testing the plate by percussion, it should be held first by the large end, about 2 or 2 ½ inches from the top and 1 inch from the left of the center joint. This will give you the C=261.63 at center and G#=207.65 at the lower perimeter. However, when the plate is turned around and held by the small end at about the same location, A=220 is obtained with quite a strong response. Therefore, the final tone response of the back plate when graduated from 10/64 of an inch at center to 6/64 of an inch at the perimeter is A=220, C=261.63, G#=207.65.

With regard to the frequency of C=261.63 noted on the back plate when final graduation is completed from 10/64 of an inch to 6/64 of an inch, this frequency seems almost unbelievable, as it will be noted that C=261.63 on the frequency list is 14/64 of an inch, and here we are receiving it at 10/64 of an inch to 6/64 of an inch on the final graduation.

We can only conclude that when the final graduation is made that in some way it vibrates on a smaller area. This is not undesirable. In fact, it is very possibly an advantage that maple has this ability, giving the plate a wider range of resonance.

The writer has an old Italian violin, a Francisco Ruggeri 1659. It, too, responds on the back plate to C#=277.18. The final complete tuning, therefore, on the graduated plates is: top spruce plate 8/64 of an inch to 7/64 of an inch. B=264.94, F=174.61 at perimeter. The bar tunes at F≐349.23. The maple back plate is A=220, G#=207.65 at perimeter, and also C=261.63, and C#=277.18. The air column is C#=277.18. The sides augment the resonance for the lowest tones on the G string: G=196, G#=207.65, A=220, A#=233.08 and B=246.94.

It should be stated here that the graduated thickness of the finished plates are not "set in cement." Other thicknesses may be used. Indeed, when we study the history of Stradivari and other old Italian instruments, we find that somewhat different thicknesses have been used, thicknesses that range for the top plate 6/64 of an inch all over to 10/64 of an inch for the back plate. Others range as high as 12/64 of an inch for the back plate. Other thicknesses are given for a number of old Italian violins at 4 millimeters to 2.5 millimeters for the top plate, to 5 millimeters to 3 millimeters for the back plate.

It should be mentioned that when Stradivari graduated his top plate to 6 1/64 inches all over, he put in a much larger bar. He made the bar 12 millimeters high, which is almost a half-inch. The heavier bar increased the frequency for the higher tones of the tuned octave.

As the years went by, it became generally believed that 8/64 of an inch for the top plate and 10/64 of an inch for the back plate gives the best balance. When the thickness is too thin, the higher tones suffer; and when too thick, the lower tones are weak.

When the violin is finished and the plates glued to the sides, the frequencies of the lowest tones on the G string reach into the sides for additional area for resonance.

When both plates are finished and graduated, top plate 8/64 of an inch to 7/64 of an inch, back plate 10/64 of an inch to 6/64 of an inch, the testing of the top plate should be done before the F holes are cut and finished tones are tested.

All of these different thicknesses indicate that the sand patterns would show slightly different patterns of formation. The vibrations have the ability to select different areas of the plates to find resonance. They will find it on plates graduated to slightly different thicknesses.

One rule that must be observed is that the plates may not be tuned in unison. They are generally tuned one tone apart. However, when we study the graduated thicknesses of the old Italians, we find that they do not always adhere to this rule. There are variable differences of the one-tone rule between the plates. They are not in unison, however.

When laying out the lines for the graduating thicknesses, the starting point should be on a line; for the top plate across the bridge line, making equal measurements up and down from the bridge line 8/64 of an inch at the center to 7/64 of an inch at the perimeter. For the back plate, the beginning center line is at the sound post line, reducing the thickness up and down from 10/64 of an inch to 6/64 of an inch at the perimeter.

When it is understood that the plates vibrate from equal areas

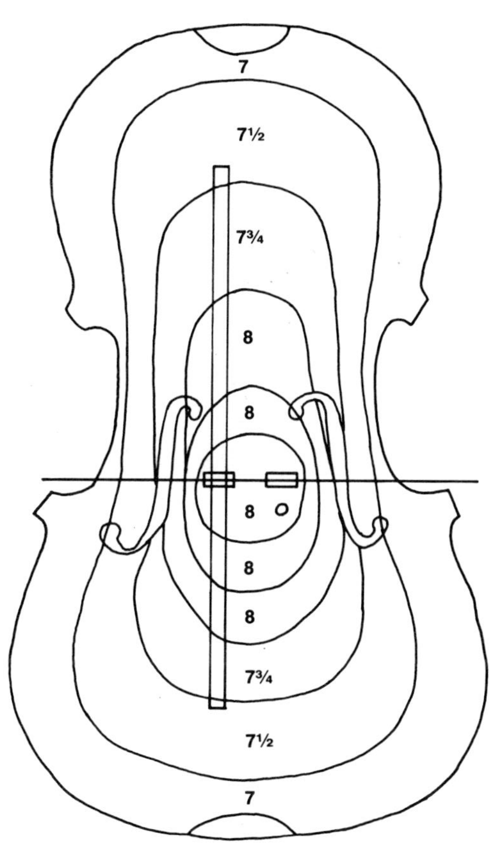

Location of graduating lines. Thickness in 1/64 THS inch.

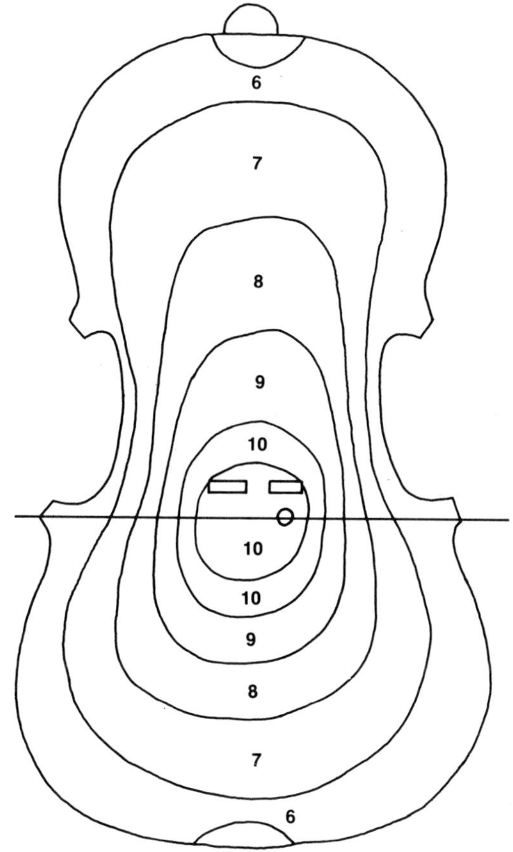

Location of graduating lines. Thickness in 1/64 THS inch.

and thickness on each side of the bridge and post lines, the truth of this graduating is realized.

Some books on the construction of the violin start the graduation from the greatest thickness at the exact center of the plates from the upper and lower ends.

The diagrams on the previous page explain this graduating.

CHAPTER 6

Sand Patterns

The writer, in various experiments, became convinced that as far as the vibration of the plates is concerned, flat plates vibrate in the same manner as arched plates. The arching has nothing to do with the vibration of the plates, but rather it has everything to do with the vibrating air column, which we will discuss later. At this time, a discussion of the vibration of the plates is in order. It was decided that with flat plates, we should be able to obtain true pictures of sand patterns on the plates. Each one-half tone was put under vibration separately—first the top plate. Then the top plate was removed and sand sprinkled inside on the back plate. A small wood piece was glued across the bridge and post area in order to take the bridge and sound post. Each frequency was bowed a number of times until the sand arranged itself into patterns, the results of which are shown here. The patterns were taken on the general plate thicknesses of 8.03 to 7 1/64 inches for the top plate and 10/64 to 6/64 of an inch for the back plate. The patterns are quite revealing. It should be noted that the plates do not vibrate the same way as they do to test the fundamental tone of the plates by percussion, which produce B = 246.94 for the top and A = 220 for the back. It should be *especially* noted that each vibration as produced by the strings selects areas of the plates and splits the plates into various sections which are in resonance with the frequency of the string.

For those who are not familiar with sand patterns, it should be explained that an equal amount of sand is sprinkled on each plate. When a tone is bowed on a string, the vibrations of the bowed tone vibrates those portions of the plate that are in resonance with the particular frequency. It will be observed on each diagram of sand patterns that there are dark areas and light areas. The dark areas are portions of the plate that are at rest. The light portions of those sections of the plate that are under vibration.

When the tone is bowed on the string the sand arranges itself into these patterns of vibration and rest.

The set of sand patterns on the thicknesses of 8 to 7 1/64 inches for the top plate and 10 to 6 inches for the back are more nearly the thicknesses used. Another set of sand patterns was made at 11.50 1/64 inches for the back, but it is very similar to the 10/64 to 6 1/64 inches.

The sand patterns were made with a bar 10 inches long, which is similar in size to those installed in most Stradivari violins. This is the size of most of the old Italian bars. We noted some Stradivari bars at 9 inches. Most of Stradivari bars were 9 ½ inches long, 13/32 of an inch and 11/32 of an inch deep, and 3/16 of an inch thick.

There is a graduating thickness as done by most violin makers that was determined, and should be corrected. Most makers, when graduating the plates, graduate their thickness at the edge to 6.5 or 7 1/64 inches, but when they reach the extreme edge, they bring the thickness up to 7, 7.5 1/64, or 8 1/64 inches at the extreme edge. This appears to be a mistake, as the vibrations of the G string reach into the sides for an additional area. Both the top and back plate do this. In order to prove this, hold the violin under the chin so that the left hand is free. Now bow the open G string and at the same time hold the left hand finger on the side of the violin. A very strong vibration will be felt. Now bow the open A string. It will be observed that no vibration is felt on the sides when the A is bowed. Therefore, the sides and linings should be no thicker than the final thickness at the edge of the plates, and it is beneficial to graduate the extreme edge of the plates to 6.5 or 7 1/64 inches. There must be a continuous equal graduation right into the sides. The writer read of one man who made a small cut all around the inside edge of his violin plates, which improved his tone. This permitted the vibration of at least the G string to reach into the sides.

The vibrations do not go uphill, so in order for them to use the sides, there must be a continuous downhill graduation at the plate edge right into the sides.

CHAPTER 7

Air Columns and Air Volume

The volume of air in the violin gives a frequency of C = 261.63, but there is a slight difference. When the sound post is installed, it must in some way change the wavelength, because when the post is installed, the frequency is raised to C# = 277.18, which is exactly where it should be. It will be noted that C# = 277.18 is exactly half way between G = 196 and G = 392. The tuning of the violin is of this lowest octave, and C# = 277.18 is halfway between this tuning. When the violin is tuned to this lowest octave, all other higher octaves will have resonance. As explained previously, the several parts of the violin are tuned to specific frequencies. The top plate, which was at first graduated to 8/64 inches to B = 246.94 equals thickness all over and is now graduated from the 8/64 inches at center to 7/64 inches at the perimeter, will still remain at B = 246.94 at center but will have a lower frequency at both ends of the plate of F = 174.61. The back plate, at first graduated to A = 220.10, 10/64 of an inch and now graduated down to 6/64 of an inch at the perimeter, will have a frequency at the center of C = 261.63 and then both ends of the plate will go down to G# = 207.65. The air column is C# = 277.18 and the bar will have a frequency of F# = 370. The sides give additional area for resonance for the lowest frequencies of the G string. These thicknesses and graduations produce the necessary conditions whereby the vibrations of the whole range of the strings can find areas to produce resonance for the whole range of the strings.

The correct height of the arching to produce the correct volume of air of C = 261.63 is 11 millimeters for each plate. This height, together with the correct height of the sides, 1¼ inch lower bout, 1³⁄₁₆ inch upper bout, produces the correct air volume, 22 millimeters for both plates.

It should be stated at this point that Savart, in testing for the frequency of the air column, obtained the wrong frequency. He obtained a frequency of C = 512 on the old pitch. This is one octave higher than the true pitch of the air column. The correct frequency is C=261.63 on our pitch of A=440.

Now we must speak about the so-called bass bar. Considerable misconception has been entertained about the bar for many years. It is not for bass tones. In fact, it adds an additional load to the top plate which makes for *higher* frequencies. When the bar is installed, the frequency of F#=370 can be detected by percussion. Hold the plate by the large end about three inches down between the thumb and finger, and *gently* strike the center of the bar with the finger. F#=370 can be detected; check with tuning fork. However, this does not mean that F#=370 is the only frequency obtained on the bar. The lowest frequencies of G 196, G#=207.65, A 220, A#=233.08 and B=246.94 use the sides for additional area. This may be observed with the tuning forks. The frequencies used by the bar start at C=261.63 and up—C#=277, D=293.66, D#=311, some of E=329, F=349, F#=370, G=392, G#=415.3, B=493.9, and F=698.46.

It is advisable to test the frequency of the bar when it is first glued to the plate. Test tone as it is cut down. A bar 10 inches long by 3/16 of an inch thick, 13/32 of an inch and 11/32 of an inch deep will test out F#=370. It will test out higher than F#=370 when first glued into the plate.

Savart made up some apparatus through which they blew to determine the air column frequency by blowing over the F hole. However, this is not necessary. It is quite simple to blow over the F hole with one's mouth. With a little practice, the tone is quite noticeable.

It should be stated here that some of the old Italian violin makers made the arching of their plates 12 millimeters high instead of 11. This extra height does not lower the frequency below C=261.63.

There are some other adjustments for the air column that should be mentioned. It has to do with the reduction of the depth of the sides from 1 ¼ inches at the lower bout to 1 3/16 inches at the upper bout. This was done to equalize the air volume of the upper and lower bouts.

It involves a complication with the volume of air contained in the arching of the plates, and it presents something of a paradox.

The old Italians, by designing their violins with a 6 ½ inch upper bout and a 8 ⅛ inch lower bout, placed the line of equal area for the upper and lower bouts at a line ½ inch below the bridge line. By so doing, the air volume contained in the arching of the upper half of the plates lengthened and increased the air more than in the lower half. By reason of this, the air in the upper halves of the plates is compounded, making it worse. Now the only way they had to equalize the air volume for the entire body of the violin was to lower the upper bout to 1 3/16 inch. This equalizes it.

On page 34, calculations for this equalization are given.

The arching of the upper plate is increased more than the maple back plate. This increases the air to compensate for the reduction of the air as occasioned by the inclusion of the bar. So it is evident that the air must be equalized for both upper half and lower half sections as with the wood area load of the plates.

Another condition should be done. The inside of the violin should be sanded quite smoothly. The smoother, the better. It allows the air volume to vibrate better. This is explained in the acoustical work by Herman Hemholtz. We might mention here that Herman Hemholtz made small spheres with a hole to fit into the ear. They were made of different sizes and the air column tuned to different frequencies so that when one was placed in the ear and the same note struck on the piano, the resonance of the sphere sounded in sympathy with the piano note.

Some sound scientists have said that the violin is essentially a Hemholtz resonator. We must say that the violin is much more than this. While the Hemholtz resonator is tuned to one frequency, the violin must give resonance to the whole range of frequencies as bowed on the strings.

Speaking of the bar, it was observed that the bar should be set on an angle with the angle of the upper to lower bout. It can be observed that there is a better response with the tuning forks E=329.63, and F=349.23 to F# =370 when placed on an angle of the upper and lower bouts rather than placing it even with the grain of the wood.

CHAPTER 8

The Bridge

Mention must be made concerning the special design of the bridge. The bridge acts as the feet of two tuning forks. Each foot conducts different vibrations. You may well ask, if the bar is for the higher frequencies, why is it on the G string side? Violin makers are aware that it is better for the construction to place the sound post under the E string side, as this is where the greatest weight is located. Therefore, the bar had to go under the G string side. This then necessitated special design for the bridge. It will be observed that the special cuts made at the center of the bridge and at each side *compels* the vibrations to proceed to the center of the bridge. Having arrived at the center of the bridge, they naturally migrate to whichever side is in resonance with each vibration.

The top plate actually vibrates with dual resonance. It splits in two halves right down the center between the bridge, as observed on the sand patterns.

These special cuts in the bridge, which compel the vibrations to proceed to the center of the bridge, also distribute the energy equally to both sides of the top plate.

CHAPTER 9

Tuning

The F holes do nothing for the resonance of the upper plate. In fact, when the holes are cut, they diminish the resonance of a number of the frequencies of the tuned octave that were there before the holes were cut, and observed with the tuning forks. However, when the bar is installed, all the lost frequencies are again restored. Not only does the bar bring back the lost frequencies, but they sound with the tuning forks right over the location on the plate where the bar is located, at the location where the left foot of the bridge rests. All this gives proof that the bar is not for bass tones.

We again mention the tuning. The back plate is tuned to A=220, the top plate B=246.94, the air column C#=277, the bar F#=349.23.

The finished violin sounded with the tuning forks is as follows:

At center of plate between the bridge feet—G=196, G#=207.05, A=220, A#=233.08, B=246.94.

Over the bar—C=261.63, C#=277, D=293.66, D#=311.13, E=329.63, F=349.23, F#=370, G=392, G#=415.3.

All across top plate at bridge line—A=440, A#=466.16, B=493.9, C=523.25, D=587, E=659.26, F=698.46.

Back plate—all forks sound well at center of plate. Some forks sound better at right on bar side.

Having tuned for the lowest octave, all other higher octaves will have resonance.

Note the similarity between the sand patterns for the lowest octave and the sand patterns of the same tones one octave higher.

With regard to the width and shape of the F holes, reference is made to the work of Herman Hemholtz, *Sensations of Tone,* in which he gives directions for his resonators. In these directions, he speaks of experiments to determine the correct size for the opening of the holes in relation to the size of the sphere, giving the best response. He regarded an opening size of ¼ the diameter of the sphere to give the best result. He also states that a narrow slit adds greater power to the tone. The area of the F holes is ¾ of an inch per hole, or 1 ½ inches for both holes, when transformed into a sphere, and the air column is also transformed into a sphere. It is found that the area of the F holes is one fourth the diameter of the air column. This indicates that the old Italian violin makers, in determining the shape and size for their F holes, arrived at the same general ratios for the size of their F holes as did Herman Hemholtz for his resonators. He also found that holes long and narrow gave greater power to the tone. The holes may not be too narrow or the tone will be sharpened too much, and if too wide, the tone is sort of bland in quality. In other words, the narrow hole concentrates the tone too much and the wide hole opens it too much. Experiments on the F hole indicate that they may not be altered in size or shape. The regular Stradivari F holes are quite correct. The writer has been unable to alter them.

When a violin attains maximum resonance due to favorable wood and correct construction and graduation of the plates, it obtains a certain richness of tone not acquired in many instruments. It is often experienced in certain old Italian violins. It is difficult to explain it, but when you have it, it is very rich and the tone comes so very easily.

THE PURFLING

The writer has had some concern regarding the purfling used today, with the black fiber laminate on the outside of the one white wood piece. When it is understood that the lower frequencies reach into the sides for additional area, it is quite probable that the black fiber laminate would dampen the resonance from going into the sides. All-wood purfling would allow more freedom for the vibrations to proceed into the sides.

CHAPTER 10

Some History on Pitch

A few observations upon the history of pitch seem pertinent, with regard to the size and load of the bar. It is quite well established that the old Italians used a shorter bar than that used by modern makers—generally, about 9 ½ inches long by 3/16 inches wide by 5/16 inches deep. Somewhere over the years, someone decided that a longer bar was necessary, as is our bar today. The reason given and accepted without exception by all writers on the subject is that the longer bar was necessary because of the raise in pitch over the years. However, when we study the history of pitch, we must have some doubts about this. In the first place, we find that the old lutes and viols of the sixteenth century were tuned to a pitch as high as A=475 and even as high as A=500. There was a high church pitch also in this range. Some authorities on the subject inform us that the pitch was lowered upon the introduction of the violin, as this instrument could not be tuned up to the high pitch of the lutes and viols. Now, somewhere in the sixteenth century, this high pitch was lowered and a number of pitches were used in different locations all the way from A=395 to A=440. A general compromise seems to have been made at A=424.2, and A=422.5 was the pitch of Handel's fork. So we may reasonably believe that the old Italians of the seventeenth century and early eighteenth century used a pitch of A=424.2 or possibly A=430.52. Fetis gives their pitch at C=512, which is A=430.52. Now, a pitch in this range is not very different from our own of A=440. A few years ago our pitch was A=435.

Therefore, when the old Italians tuned their instruments to a pitch of A=430.52 or A=424.2 and installed a bar 9 ½ inches long, what reason do we have now to install a bar 11 inches long when our pitch was A=435 or A=440 now? The question therefore presents itself as to when and how this larger bar came about. There is a definite and plausible answer to this, although a mistaken one. About the year 1850, the pitch was suddenly increased again and was raised from the older standard of A=424.2 to A=452 and even A=460. Now, it is quite evident that someone at this period decided that with this increase in pitch, the violins needed a longer bar, which in itself was a mistake. Now, however, that the pitch is back again very close to the pitch of A=424.2, as in the old Italian violin making days, what possible reason can we have for a longer bar than they used? The question therefore is, should we install the old Italian 9 ½ inch bar? The writer in no way suggests that this should be done. No experiments have been conducted to prove that the 9 ½ inch bar is better for the tone, except that it might be said that the 11 inch bar might interfere with the resonance of the upper and lower ends of the top plate.

HISTORY OF PITCH

The reader is referred to the *Encyclopaedia Britannica* on the history of musical pitch, in which it refers to the pitch being lowered upon the introduction of the violin, because the violin could not be tuned up to the high pitch of the lutes and viols, which was at that time A=475 and A=500.

It is possible that at the time of the old Italian violin making, the pitch could have been A=424.2 and A=430.52, as stated by Fetis. Herman Hemholtz gives a complete history of pitch in his book *Sensations of Tone,* in which there are indications that A=424.2 was used at the time of the old Italian violin making. The old pitch of C=512, A=430.52 could also have been used at some time of the old Italian making days. Hemholtz, in his history of pitch, gives the pitch of A=424.2 and others at A=420, and A=423 as the mean pitch for two centuries all over Europe.

Graduating with a pitch of A=422 or A=435 would have little difference to the resonance of the areas. The sand patterns might be slightly different. An instrument graduated to the thicknesses with a bar giving certain resonance response based on the thicknesses for a pitch of A=430 would have exactly as good resonance to an instrument tuned on the basis of A=440 or A=435. It simply means that the plates would vibrate with a slightly different formation of sand patterns.

There is a tuning fork response on some top spruce plates when graduating the inside thicknesses that should be mentioned. On certain plates, when graduating the inside, the tuning forks can sound very strong on the plate all the way from C=261.63 to C=523.25, even before the graduating is completed. It must be the result of some particular quality in the wood. When this occurs, you may be sure that your finished violin will have exceptional tone. The writer wishes he could determine some test to know this quality, other than the twenty-centimeter rods, but no test other than the tuning forks on the plate when graduating it has been found.

CHAPTER 11

The Varnish

There are good varnishes and coloring for them available on the market. The writer has not formulated many types of gums. Damar makes a reasonably good varnish for a cold cut finish and is the most easy gum to use. Do not use mastic. Damar should be cut with best grade artist's rectified turpentine and mixed with alkali-refined linseed oil. Alkali-refined linseed oil may be purchased from one of the paint manufacturers. A gallon would be enough to varnish all the violins one makes.

There is one observation that should be mentioned and that is the preparation of the wood before the varnish is applied. A sealer of some kind *must positively* be applied to the wood. The application of linseed oil varnish to the bare wood will surely *dampen* the resonance, especially on the spruce top. Hide glue thinned out quite well makes a good sealer. The glue should be thinned to a point when a drop dipped onto newspaper will soak through the paper. Glue, when sanded down smoothly, makes a wonderfully smooth surface for the application of the varnish. It is the writer's opinion that a glue size was used by the old Italians.

For coloring matter, some of the transparent alizarine colors and gold ochre may be used. Also brown madder.

DAMAR VARNISH

25 cc Damar
50 cc rectified artist's turpentine
10 or 12 cc alkali-refined linseed oil
4 cc stand oil
10 or 12 drops drier
Base coat with thinned hide glue. Sand and color with transparent yellow alizarine or transparent gold ochre; dissolve the color in

turpentine, transparent, red, or other color alizarine. Strain through fine cloth.
Add 10 or 12 cc alkali-refined linseed oil. Add 4 cc stand oil. Add 10 or 12 drops drier.
After violin is varnished with colored varnish, apply one coat of clear varnish, no color.

Some students of the old Italians believe that they used some olive oil, but olive oil is extremely plastic, and when used, it is difficult to dry the varnish. Glycerine was first discovered in olive oil. It imparts a very good leveling effect. Use not more than a few drops and omit the stand oil.

A dropper should be used to add the olive oil. Do not depend on pouring some into the varnish. When used, omit the stand oil. It is said of glycerine that it never dries.

Mr. Henry F. Payne, who was a chemist for one of the chemical companies and who has published books on varnish formulations, very kindly suggested that Kauri or Pontianak gums be used. However, they must be cooked and blended with linseed oil. The writer hopes to experiment with these gums at some time.

The ideal quality for varnish is that it be plastic. It should not chip or show a white mark when pressed with a hard object. The writer has seen old Italian chests in museums that were at least two or three hundred years old. The varnish on them looked to be the same as that on some old Italian violins in both color and texture. The writer tested the varnish on these chests with a coin, pressing it across the surface. It was quite plastic in that it bent in and would not scratch. It had every appearance of being the same varnish as seen on many of the old Italian violins. One is compelled to think that such varnish must have contained some olive oil.

The Damar varnish, as given here, is quite reasonably plastic without the olive oil.

CHAPTER 12

More About Frequency

The frequency of the air column may be determined mathematically, by multiplying the width by 5, the depth by 4 and the length by 3. Divide the total into 494,000.

Width	150 mm	average of the three widths.
Depth	37.4 mm	sides and arching.
Length	330 mm	less area of end blocks.
	113	cubic inches

$$150 \times 5 = 750 \qquad 37.4 \times 4 = 149.6 \qquad 330 \times 3 = 990$$

$$750 + 149 + 990 = 1889$$

$$494{,}000 \div 1889$$

Frequency C 261.51

Calculation based on violin widths of: upper bout 6 ½ inches, center bout 4 ¼ inches, lower bout 8 ⅛ inches.

RELATION OF F HOLES TO AIR VOLUME

F holes one-quarter the diameter of air volume. Air volume measurement by one-eighth inch squares, less area of side linings.

Surface area 4880 ⅛ inch squares

Depth of sides

$$\begin{array}{r l} & \times\ \ 9.76 \\ \div\ 512 \,\lfloor & 47628.80 \\ & 93.00 \text{ cubic inches} \\ \text{Area of plates arching } + & 20.00 \text{ cubic inches} \\ \text{Volume:} & 113.00 \text{ cubic inches} \end{array}$$

Transferred into sphere with diameter of 6 inches:

$\div\ 4 \lfloor 6.00$ → 1.50

96 ⅛ inch squares both holes equals 1.50 square inches

Area of F holes 1.50 square inches

F holes one-quarter the diameter of air volume.

The F holes are seventy millimeters long from point to point, based on Stradivari F holes. Some books on violin making give directions for F holes slightly longer, namely seventy-three millimeters from point to point, in which case the area of the F holes is slightly more than one-fourth the diameter of the air volume.

FREQUENCY OF AREA OF SIDES

The area of the sides should be considered as an extension of the area of the plates. There is a total of 36 square inches of area in the sides.

Area of sides 30 inches long
× 1.20 depth
36.00 total

Area of plates to sides—79.00 sq. inches, equals C=261.63.
Sides + 36.00 aprox ½ of plates
115.00 sq. inches total

Each half-tone is equal to six square inches added to the area of the plates—36 inches. 36 inches lowers the frequencies by six half-tones. The top plate C = 261.63 is lowered to F# = 185, which is one-half tone below the lowest tone of G = 196.

The area of the back plate is 79 square inches, which is to the sides, and gives a frequency of G# = 207. Six half-tones lower brings the back plate down to D# = 146, which is four half-tones below G = 196.

It can be seen by this that the lowest frequencies on the G string use the sides for resonance. Some resonance for the lowest tones are obtained on the plates, but the sides augment the resonance and improve it. It is important, therefore, that the perimeter where the plates join the sides be a continuous reduction in graduated thickness right into the sides.

When the thickness of the plates is finished at 7/64 inches at the perimeter for the top plate and 6/64 of an inch for the back plate, then the sides and linings should be a total of 6/64 of an inch for the back plate and 7/64 of an inch for the top plate.

Plate and body air volume balance ½ inch below bridge line:

Air volume upper body		2440 ⅛ inch squares
Depth of sides		× 9.6 inch squares
Cubic inches	÷	512 | 234240
		45.75 cubic inches
Air volume upper half plates		10.74 cubic inches
		56.49 cubic inches

Air volume lower body		2440 ⅛ inch squares
Depth of sides		9.9 inch squares
	÷	512 | 241560
		47.17 cubic inches
Air volume lower half plates		9.36 cubic inches
		56.53
		56.49
		113.02 total cubic inches

The air volume of the arching of the plates must be taken by sand measurement, in millimeters, which is more accurate:

Air upper half of plates, box, 40 × 40 × 55 mm—8800 cu. mm

$$\div\ 16387\ \underline{|\ 8800\ }$$

5.37 cubic inches
× 2
10.74 cubic inches

Air lower half of plates, box, 40 × 40 × 48 mm—76800 cu. mm

$$\div\ 16387\ \underline{|\ 176800\ }$$

4.68 cubic inches
× 2
9.36 cubic inches

10.74
9.36
20.10 total cubic inches of both plates

Calculation of the body of the violin was done by inch squares. The air volume of the arching of the plates was done by sand measurement in millimeters. The total has been transferred into cubic inches.

Included here is a chart containing the notes of the musical scale from G = 196 to C = 1046.5 together with the frequency numbers for each note. These are the total of notes and frequency numbers mentioned in this book. Reference may be made to the chart for any frequency number and location on the musical scale when reading the text.

Reference may also be made to the frequency numbers on the illustrations of the sand patterns. Every sand pattern has the frequency number at the top corner of the illustration. The reader may determine exactly on the musical chart what note is illustrated.

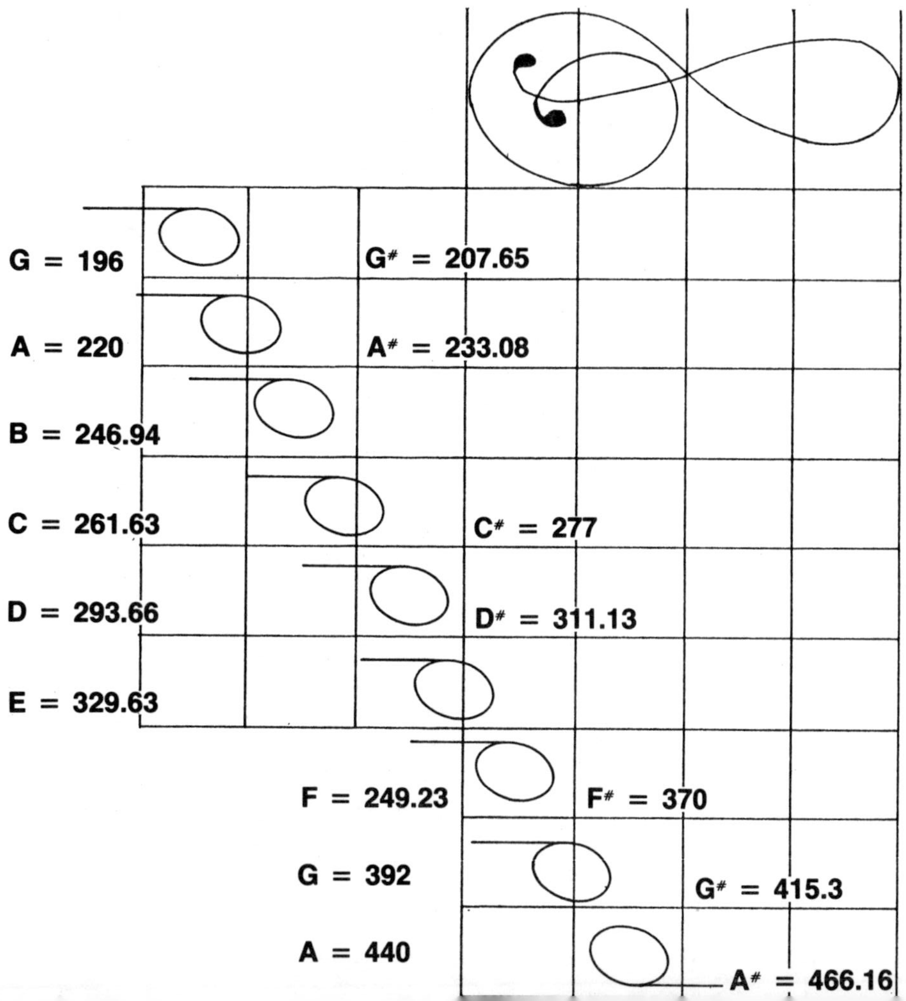
G = 196
A = 220
B = 246.94
C = 261.63
D = 293.66
E = 329.63
F = 249.23
G = 392
A = 440
G# = 207.65
A# = 233.08
C# = 277
D# = 311.13
F# = 370
G# = 415.3
A# = 466.16

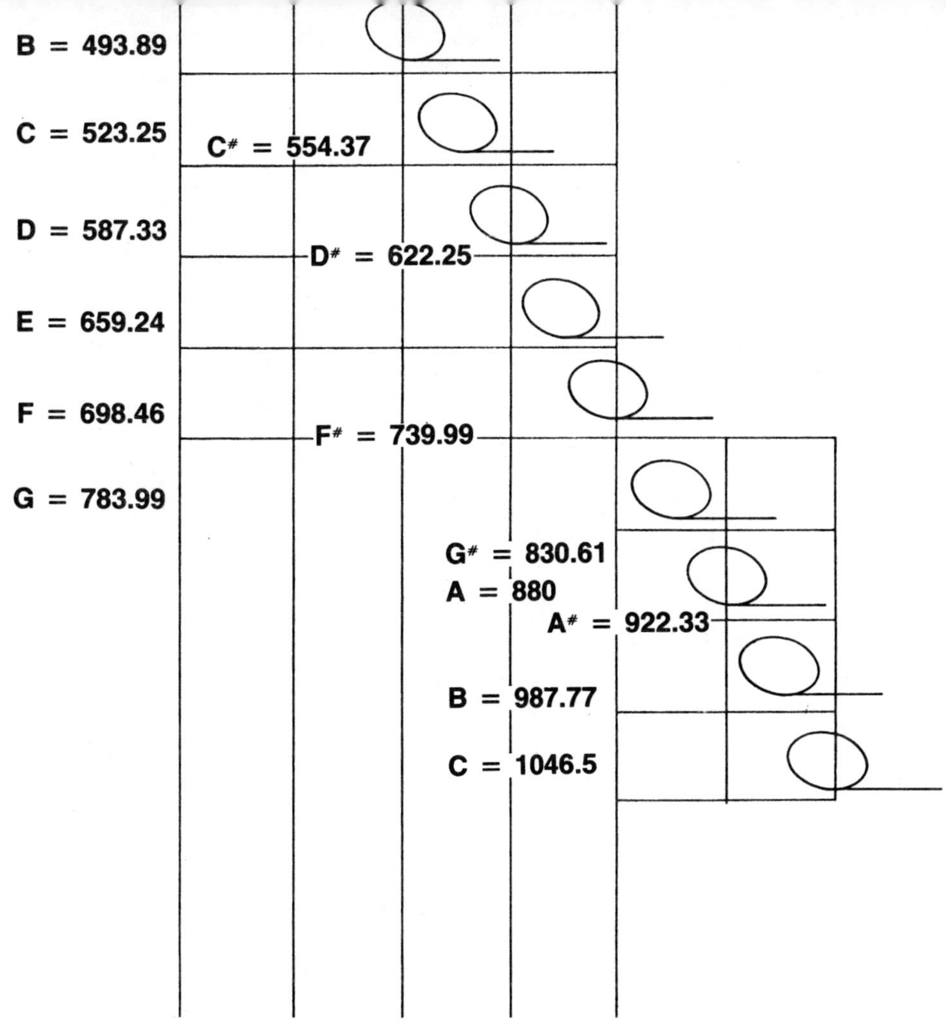

B = 493.89
C = 523.25
C# = 554.37
D = 587.33
D# = 622.25
E = 659.24
F = 698.46
F# = 739.99
G = 783.99
G# = 830.61
A = 880
A# = 922.33
B = 987.77
C = 1046.5

SOME OBSERVATIONS ON THE DEVELOPMENT OF THE VIOLIN

During the research on the scientific development of the violin and the tuning process, the writer has been amazed at the many tuning adjustments necessary to obtain the maximum resonance. It has been a source of complete astonishment at many of the tuning and scientific adjustments made by the old Italians: the construction of the body to obtain the exact air capacity so that the air volume would produce C# = 277.18 when the post was installed; the installation of the bar to give resonance to the frequencies of the higher tones of the tuned octave; the angle of the bar; the design of the instrument to give the exact area above the bridge as below the bridge, et cetera. We could go on with many more.

One writer observed that more thought went into the violin than the invention of the steam engine.

When one thinks about all this, we wonder if the artisans who constructed the instruments had some help in the scientific experimentation. The old Italians were most wonderful craftsmen, but where did they find the time for all the most delicate research?

We are told that in those days the monasteries were the laboratories. Monks did many experiments on formulations of varnish and other experiments. The writer wonders, if some of these monks, in the quietness of the monasteries, did some of the acoustical research on plates, et cetera, as the writer has done and then passed information on to the violin makers.

It takes great quietness to research the acoustics of the violin. Certainly the monks had this necessary quiet. The tones on the plates and air column must be made in extreme quietness. It has taken the writer a number of years to produce findings as written here, with the conviction that these findings are the same as determined by the old Italians.

SOME TRUTHS ABOUT STRADIVARI

Much has been said about the secrets that enabled Stradivari to produce better violins. We hear much talk about his secret varnish that is supposed to improve the tone of the instruments he made.

The truth is that Stradivari had no secrets, regarding tone or varnish. The scientific principle for the acoustics of the violin were determined before he was born. Stradvari's genius was in that he was an exceptional craftsman. However, he followed the acoustical principles discovered years before he started making violins. It is the writer's opinion that others did experimental research work and passed the results on to the violin makers, The monasteries were the laboratories in those days; The monks did much research and experimental work on varnish, and it is quite possible that they did much acoustical research on violin making.

With regard to the varnish, the writer became acquainted with an old italian whose ancestors came from Cremona. They told him that one varnish making plant made varnish for all the violin makers in that area. They all purchased their varnish from the same maker. The varnish changed in the latter part of the 18th century. More like our modern varnish, it evidently dried faster. The earlier Italian varnish probably had some olive oil in it for its plastic qualities; however, it must have taken a very long time to dry, possibly several months. The most important thing with varnish is that a sealer be put on the wood before the varnish is applied. A hide glue size is very good, as it is more flexible. If linseed oil varnish is put on bare wood it will surely dampen the resonance. It is the writer's opinion that the old Italian makers used a glue size. When wood is sealed well, most any varnish is suitable. It should be plastic, however.

Many of our modern violin makers are quite as good craftsmen as Stradivari. A knowledge of acoustics is all they need to make them equal to the old Italians and possibly better. The acoustic principles as explained in this work must be the same or very similar to that known by the old Italians.

FREQUENCY THICKNESS LIST

Based on pitch of A = 430 equals C = 512
This is the pitch said to be used by the old Italians (1600–1700).
Spruce rectangular plate, 14 inches × 6.22 inches, 87 square inches. The same area as violin plates. Total area of plates including end blocks and corners.

Divisor	Frequency	1/64″	mm
39.34	G = 191.78	4.87	1.93
37.13	G# = 203.18	5.47	2.17
35.05	A = 215.26	6.14	2.43
33.08	A# = 228.06	6.89	2.73
31.22	B = 241.63	7.74	3.07
29.47	C = 256.	8.68	3.44
27.82	C# = 271.22	9.74	3.86
26.26	D = 287.34	10.94	4.34
24.79	D# = 304.44	12.28	4.87
23.40	E = 322.54	13.78	5.46
22.09	F = 341.73	15.46	6.13
20.85	F# = 362.06	17.40	6.90
19.68	G = 383.56	19.48	7.30

Based on pitch of A = 430 equals C = 512.
Maple Rectangular plate, 14 × 6.22 inches, 87 square inches. Same area as violin plate. Total area of plates including end blocks and corners.

Divisor	Frequency	1/64″	mm
27.00	F# = 181.02	6.70	2.66
25.48	G = 191.78	7.52	2.98
24.05	G# = 203.18	8.44	3.34
22.70	A = 215.26	9.48	3.76
21.43	A# = 228.06	10.64	4.22
20.23	B = 241.63	11.94	4.73
19.09	C = 256.	13.41	5.32
18.02	C# = 271.22	15.05	5.97
17.01	D = 287.34	16.89	6.70
16.05	D# = 304.43	18.96	7.52
15.15	E = 322.54	21.28	8.44
14.30	F = 341.73	23.89	9.48
13.50	F#362.04	26.81	10.63

These are the thicknesses for each half-tone frequency. Divide frequency by divisor. The result is the thickness in 1/64 inches for each half-tone. Divide the 1/64 inch thickness by 2.52 for millimeters.

The reader will note that a plate of a given thickness when increased to four times its thickness will have a tone exactly one octave higher.

FREQUENCY THICKNESS LIST

Based on our pitch of A = 440

Spruce rectangular plate 14 inches × 6.22 inches—87 square inches. Same area as violin plates. Total area of plates, including end blocks.

Divisor	Frequency	1/64″	mm
39.20	G = 196	5.00	1.98
37.00	G# = 207.65	5.61	2.22
34.91	A = 220	6.30	2.50
33.10	A# = 233.08	7.04	2.79
30.85	B = 246.94	8.00	3.17
29.45	C = 261.63	8.88	3.52
27.77	C# = 277.18	9.98	3.96
26.20	D = 293.66	11.02	4.44
24.73	D# = 311.13	12.58	4.99
23.32	E = 329.63	14.13	5.60
22.01	F = 349.23	15.86	6.29
20.77	F# = 370.	17.81	7.06
19.60	G = 392.	20.00	7.93

These are the thicknesses for each half-tone frequency. Divide frequency by divisor. The result is the thickness for each half tone in 1/64 inch. For millimeters divide the 1/64 inch thickness by 2.52.

The reader will note that a plate of a given thickness when increased to four times its thickness will give a tone exactly one octave higher—5/64 inch to 20/64 inch; 7/64 inch to 28/64 inch.

FREQUENCY THICKNESS LIST

Based on our pitch of A = 440

Maple rectangular plate 14 inches × 6.22 inches—87 square inches. Same area as violin plates, 88 square inches. Total area of plates, including end blocks.

Divisor	Frequency	1/64″	mm
26.40	F# = 185.	7.00	2.77
24.93	G = 196.	7.86	3.11
23.54	G# = 207.65	8.82	3.50
22.00	A = 220.	10.00	3.96
20.97	A# = 233.08	11.11	4.40
19.80	B = 246.94	12.47	4.94
18.68	C = 261.63	14.00	5.55
17.57	C# = 277.18	15.77	6.25
16.63	D = 293.66	17.65	7.00
15.70	D# = 311.13	19.81	7.86
14.82	E = 329.63	22.24	8.82
13.99	F = 349.23	24.96	9.90
13.21	F# = 370.	28.00	11.11

These are the thicknesses for each half-tone frequency. Divide frequency by divisor. The result is the thickness for each half-tone in 1/64 of an inch. For millimeters, divide the 1/64 of an inch thickness by 2.52.

The reader will note that a plate of a given thickness, when increased to four times its thickness, will give a tone exactly one octave higher.

Included here are thirty sand patterns giving pictures of each half-tone frequency as produced by bowing on the strings. Every frequency produces a different pattern on the plates. It is believed that this is the first time in history that visual patterns of the vibrating plates have been produced. The dark portions are areas at rest, while the white portions are the areas under vibration. Note how each half-tone pattern changes from half-tone to half-tone. Thus, each individual frequency selects its own portion of wood to produce resonance for the frequency.

Included with these sand patterns is a group of six patterns made on a spruce plate of greater thickness, namely, 13/64 inches at the center to 7/64 inches at the perimeter. This was done to observe any difference in patterns. A great difference will be observed. In this 13/64 inch to 7/64 inch plate, it will be observed that on the lowest frequencies (G = 196, G#, A, A#) the whole width of the plate is vibrating as a unit and clear into the sides. By the time it gets to B = 246.94, it is beginning to split the plate as observed in higher frequencies. It will be noted that the top plate vibrates differently than the back. The top splits into two halves right down the middle and between the bridge. In the 13/64 to 7/64 inch plate, the lower frequencies must take the whole plate as a unit and also reach into the sides for resonance. It is in no way suggested that this thickness of 13/64 to 7/64 inches be used in graduating the top plate. These patterns are included just for reference.

Included here are six sand patterns made from a spruce top plate, namely, 13/64" inch at center to 7/64 inch at perimeter. It was done to determine what difference there might be to the vibrating plates. There is a great difference, as can be readily seen on the sand patterns. The lowest frequencies of G=196, G#207, A=220, A#=233, and B=246 are compelled to take the whole width of the plate and clear into the sides for area to obtain resonance. Violins with plates this thick would not have a good tone, especially in the lowest frequencies.

1

G = 196

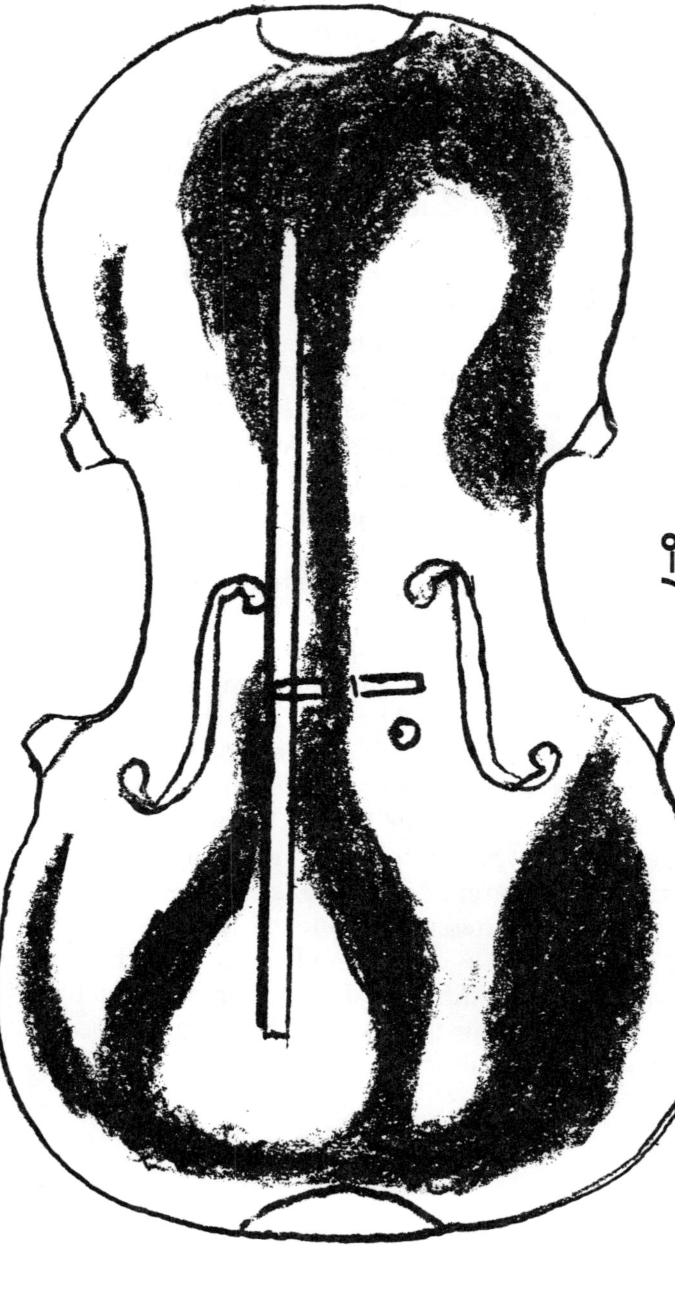

1 G = 196

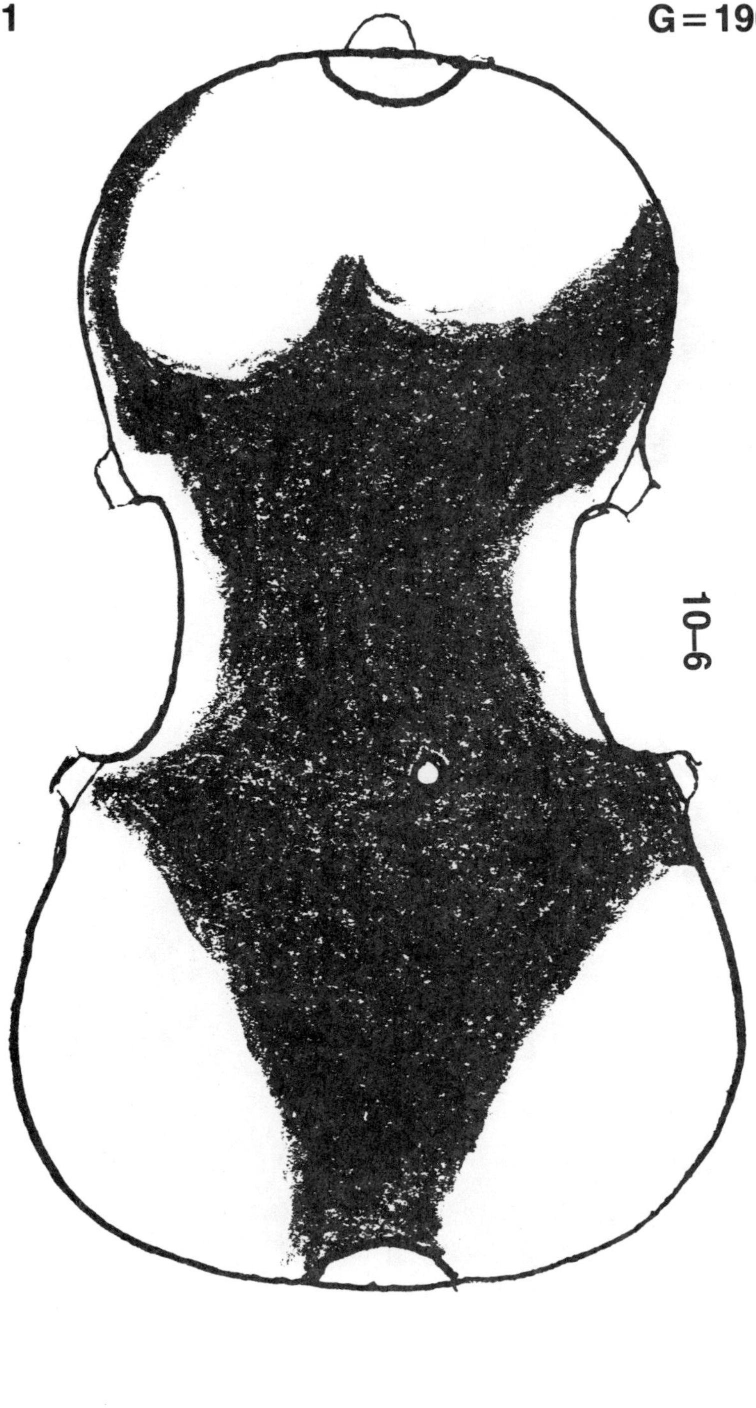

10–6

2 $G^{\#} = 207.65$

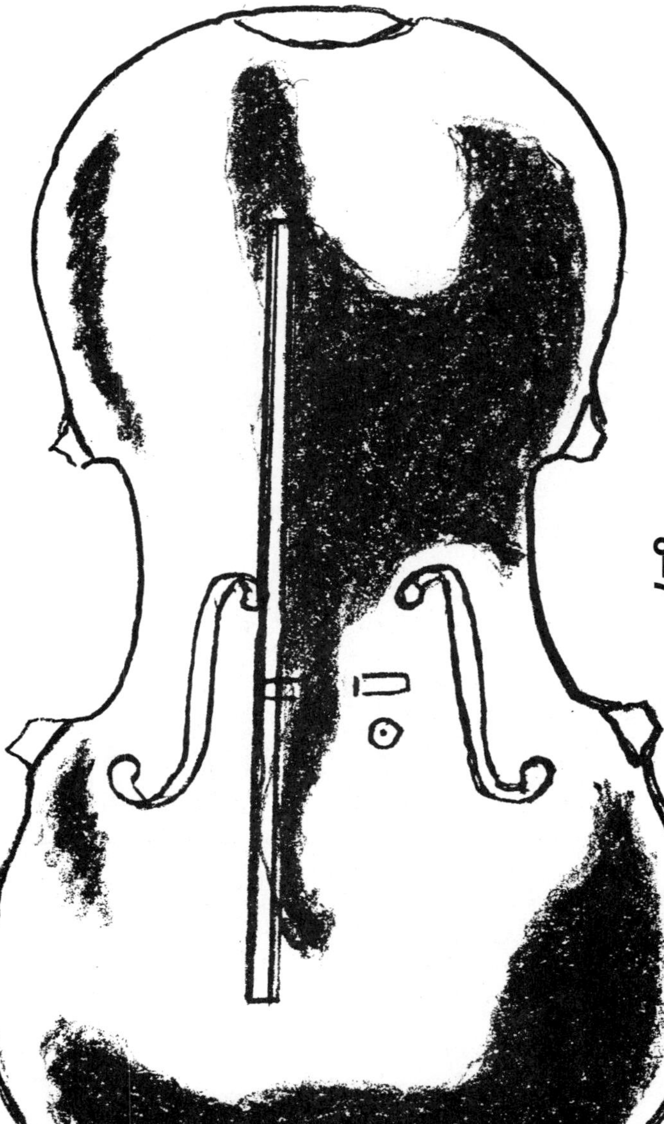

8–7

2

$G^{\#} = 207.65$

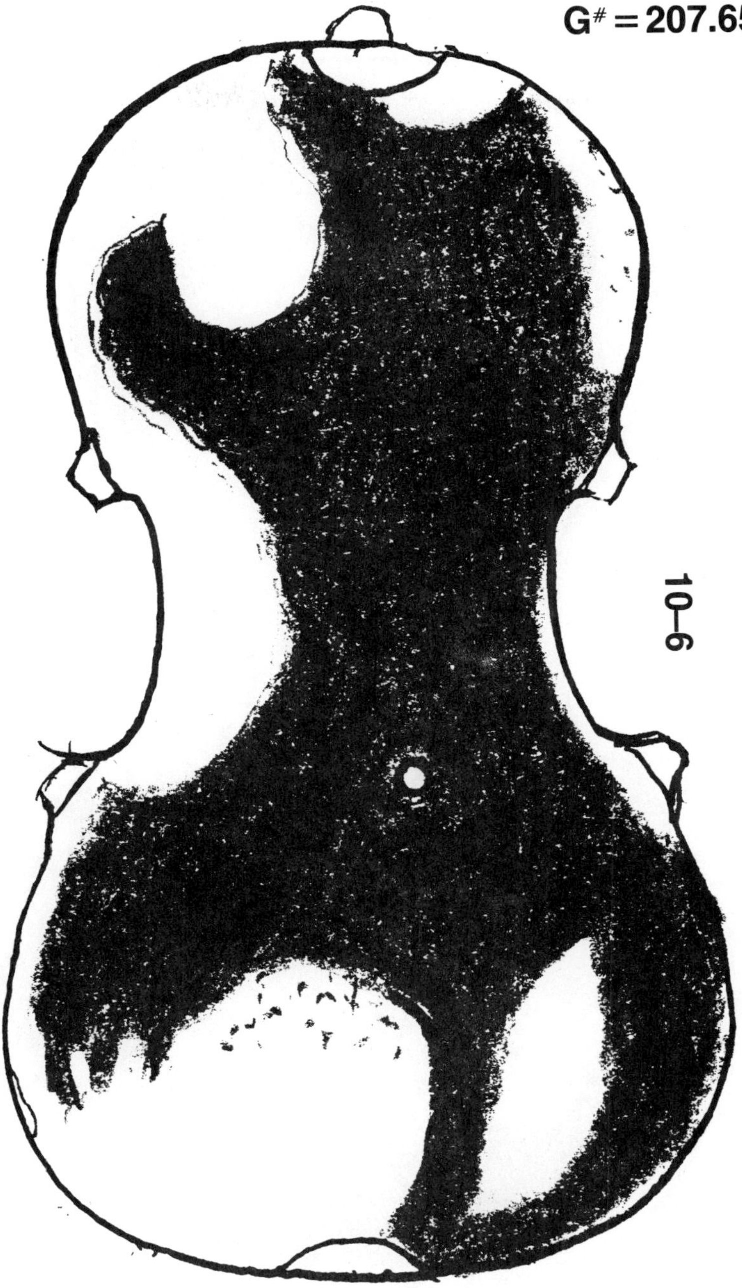

10–6

3

A = 220

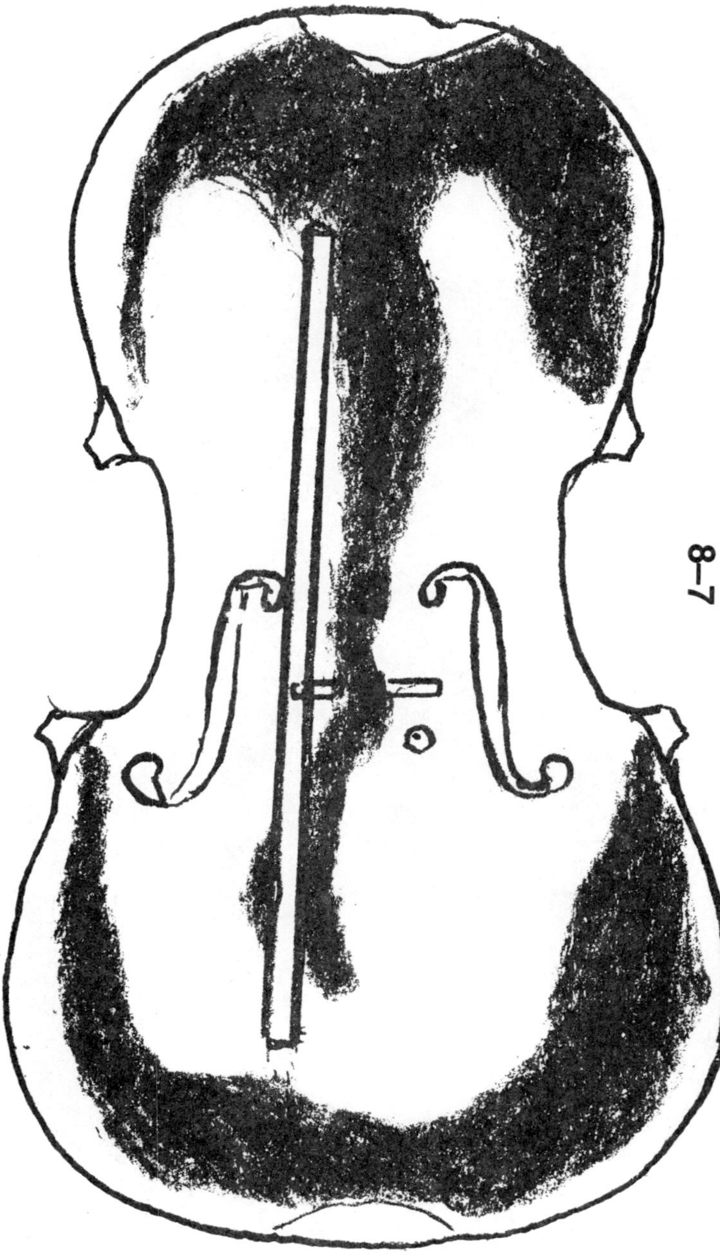

8–7

3

A = 220

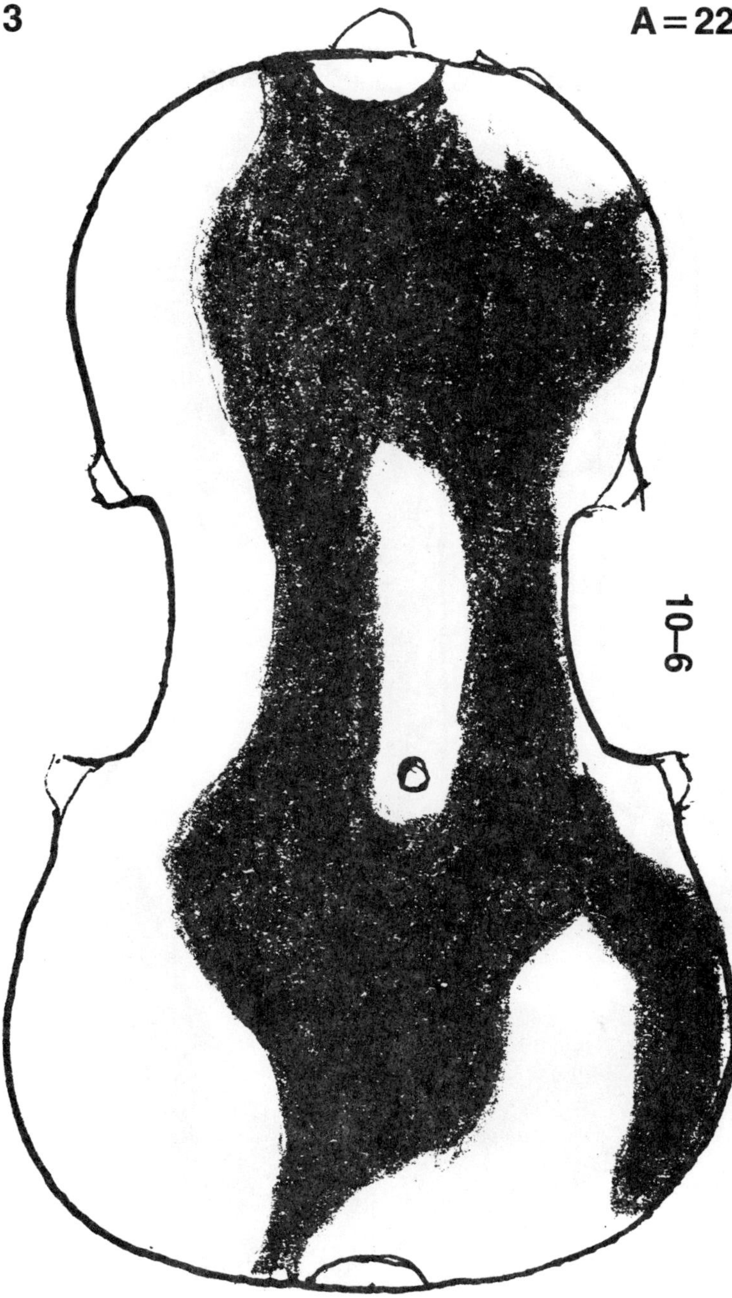

4

$A^{\#} = 233.08$

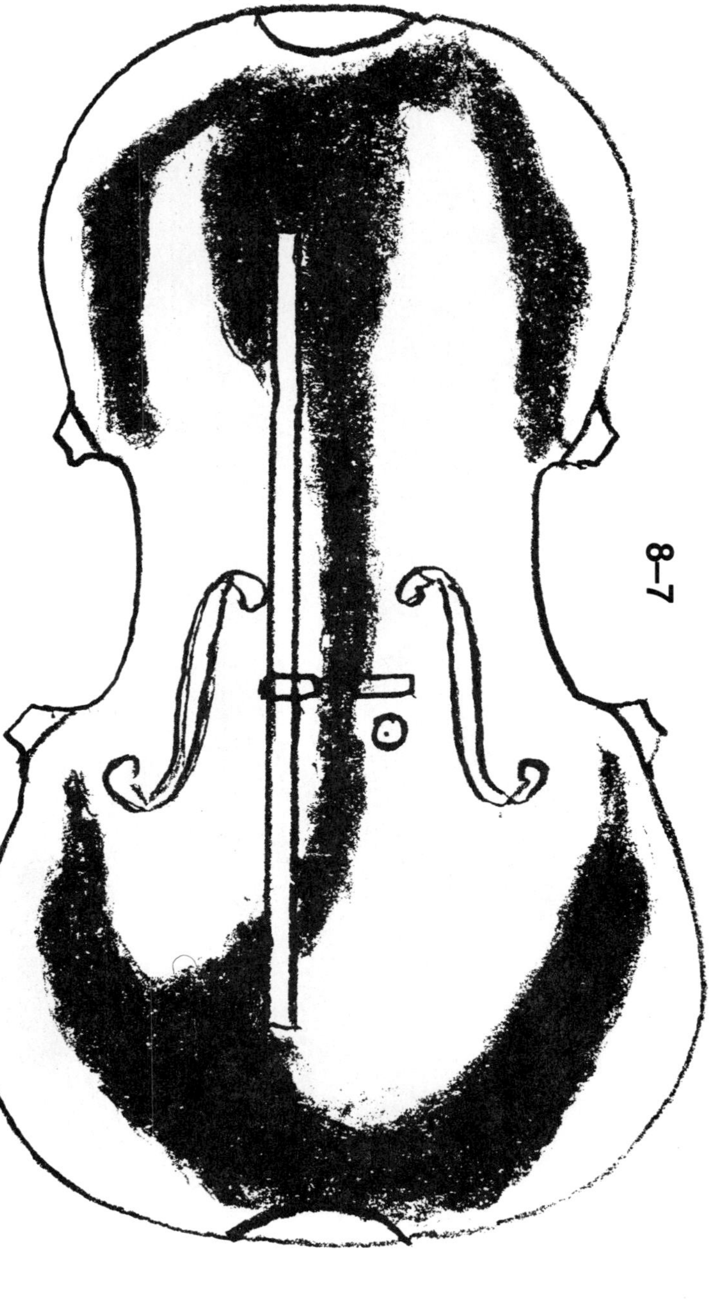

8–7

4

$A^{\#} = 233.08$

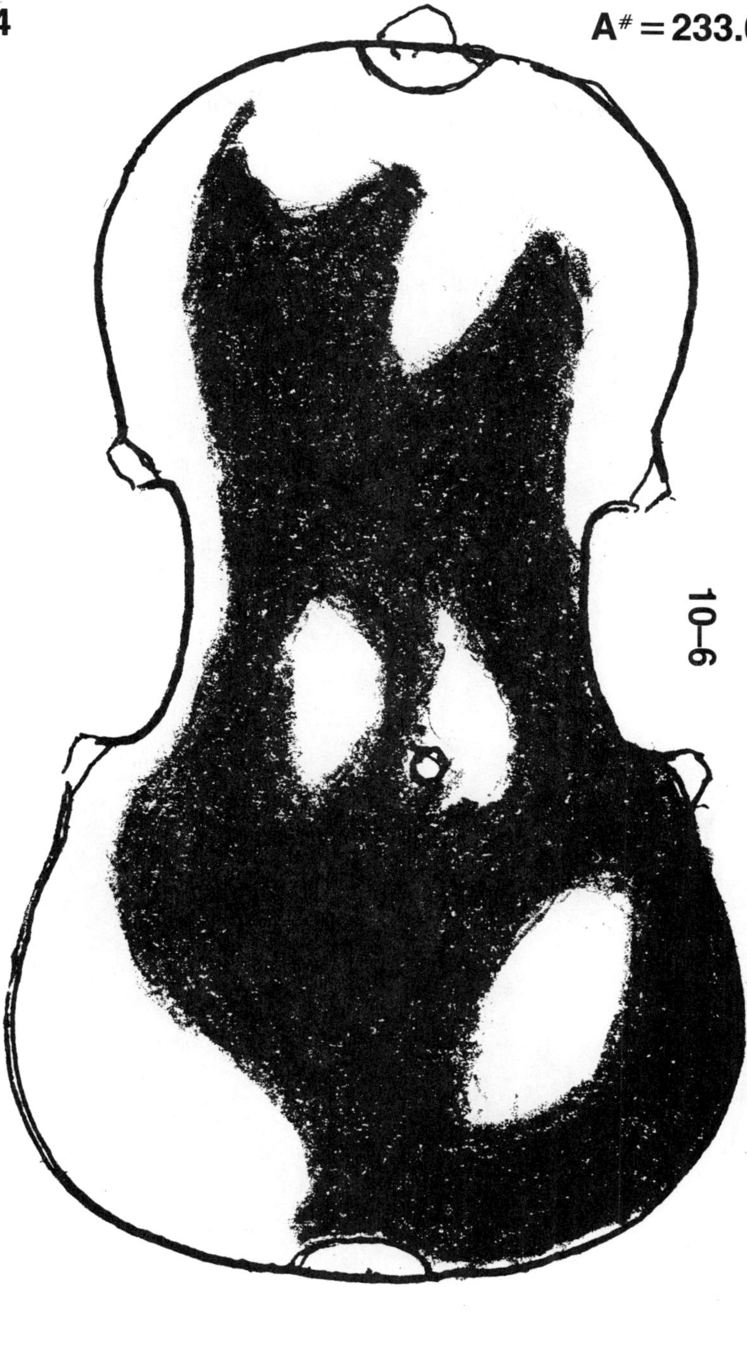

10–6

5 B = 246.94

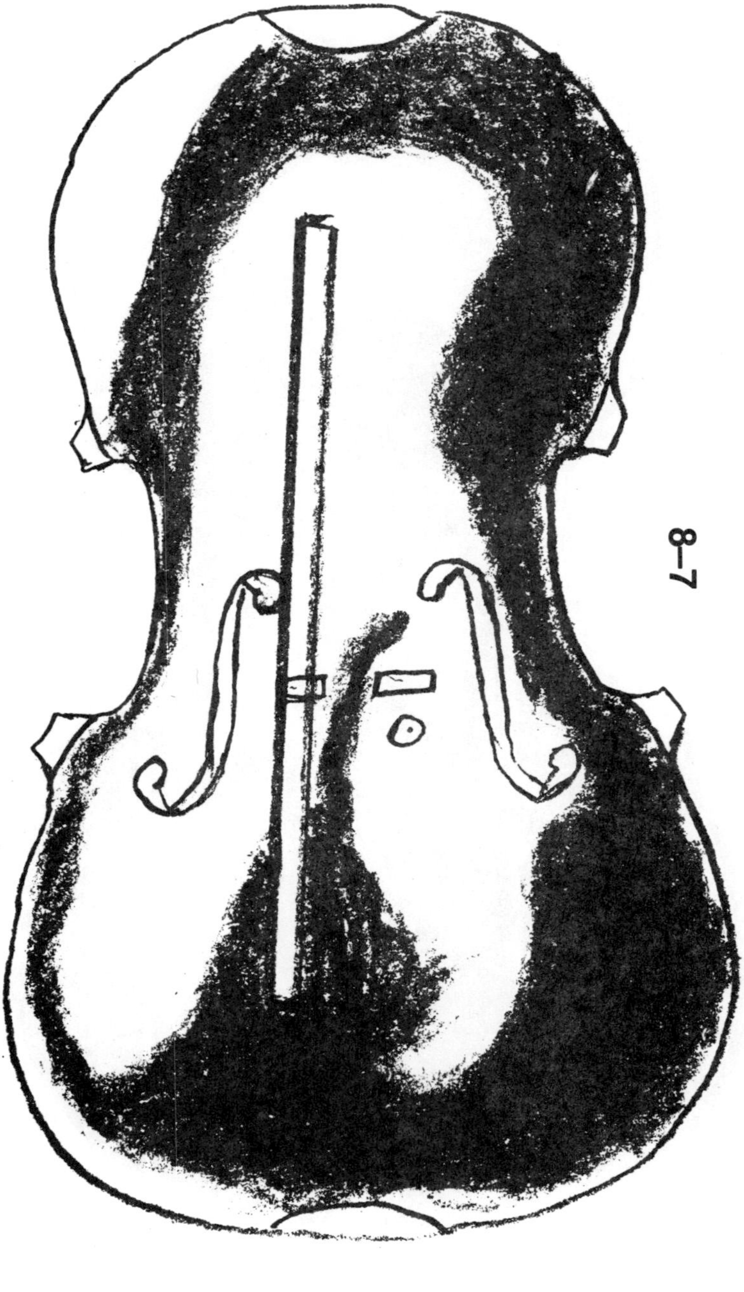

8–7

5

B = 246.94

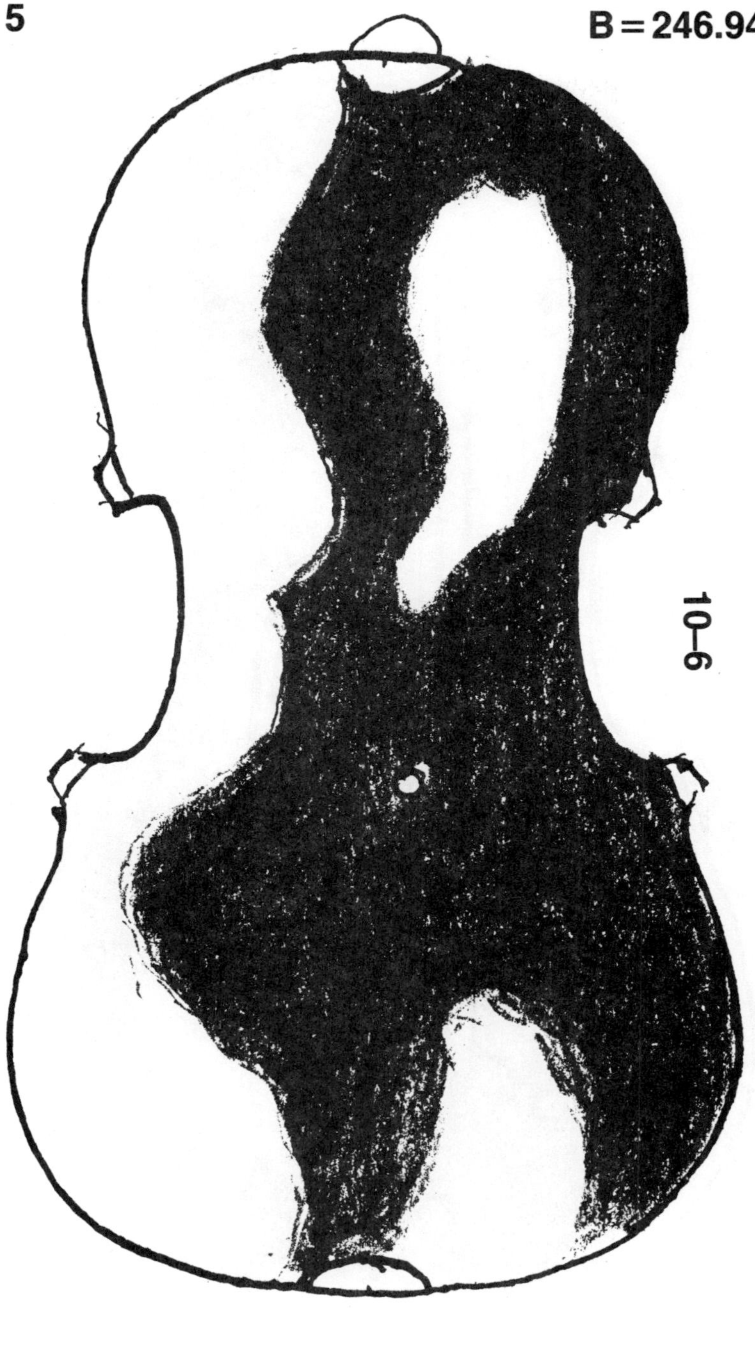

10–6

6

C = 261.63

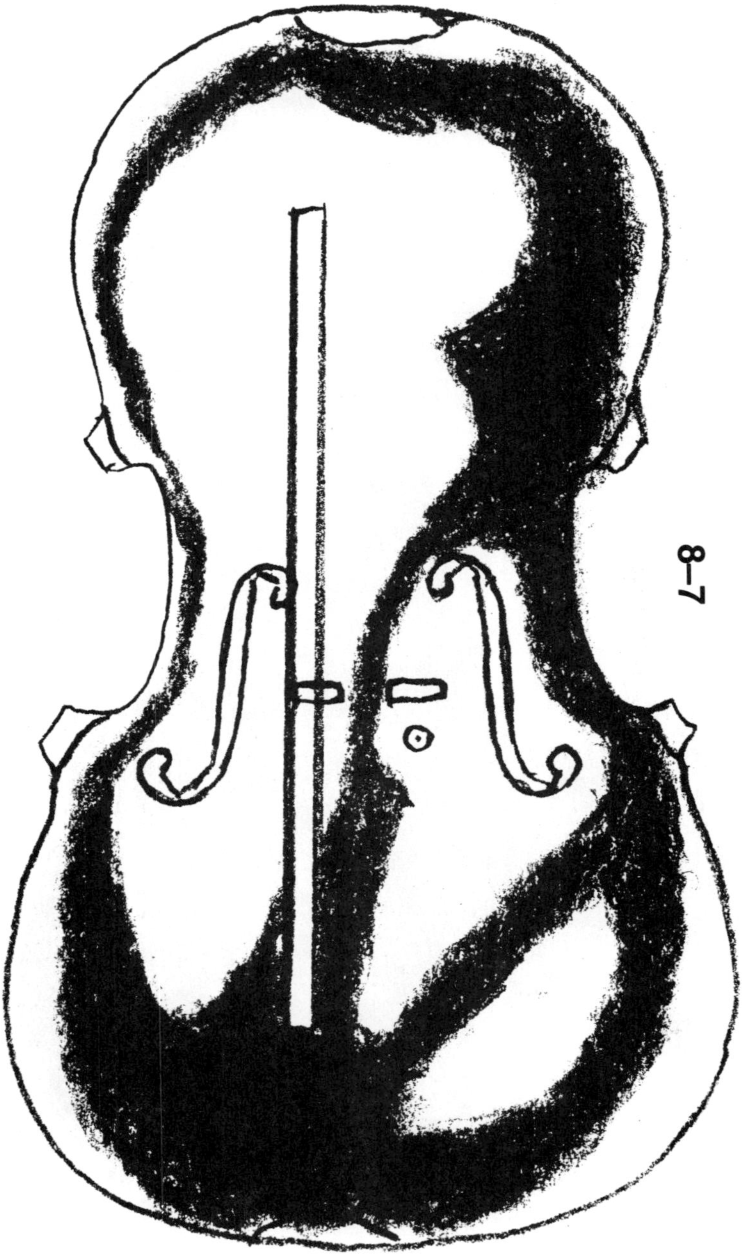

8–7

6

C = 261.63

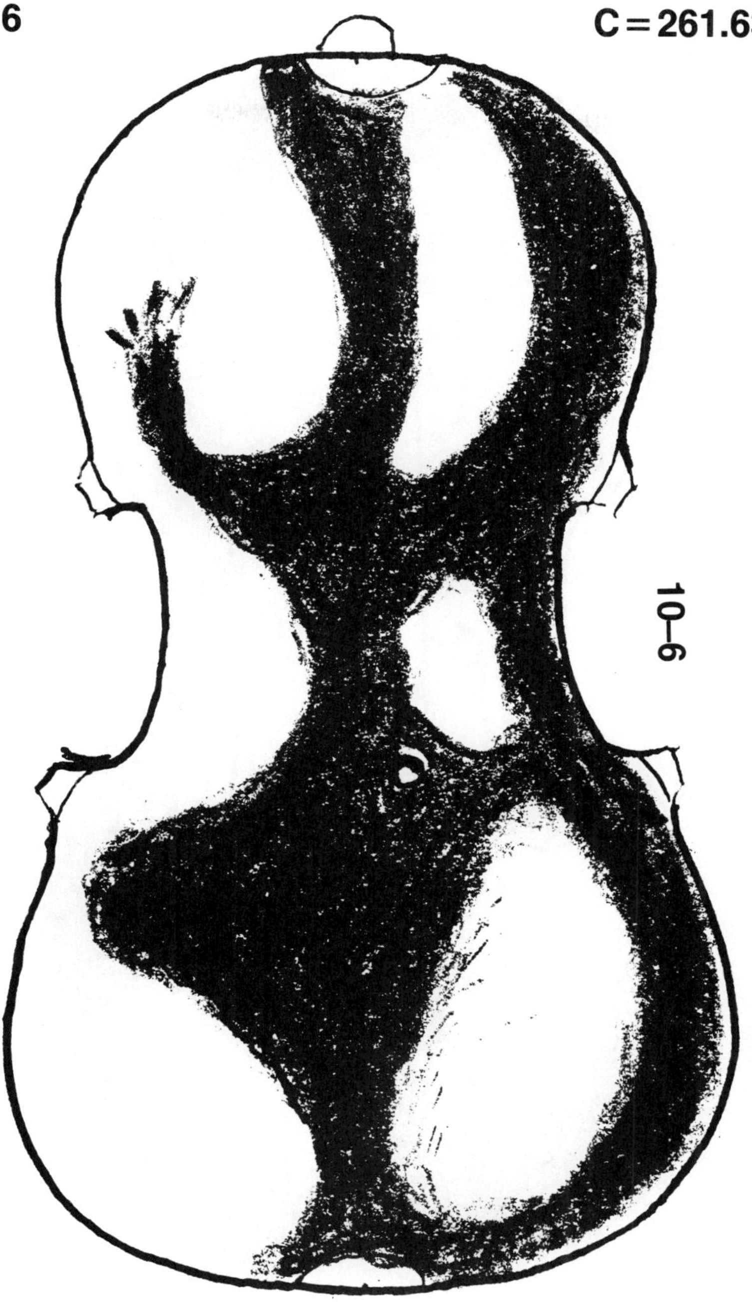

10–6

7

$C^{\#} = 277.18$

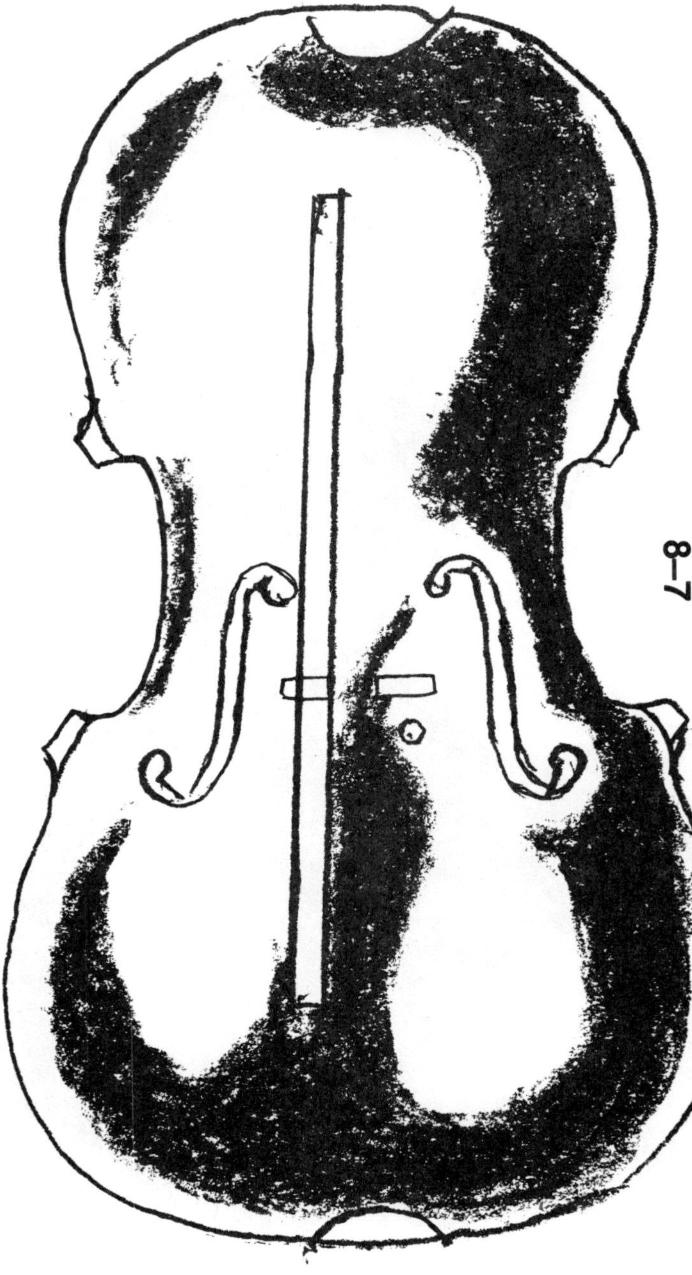

8–7

7

$C^{\#} = 277.18$

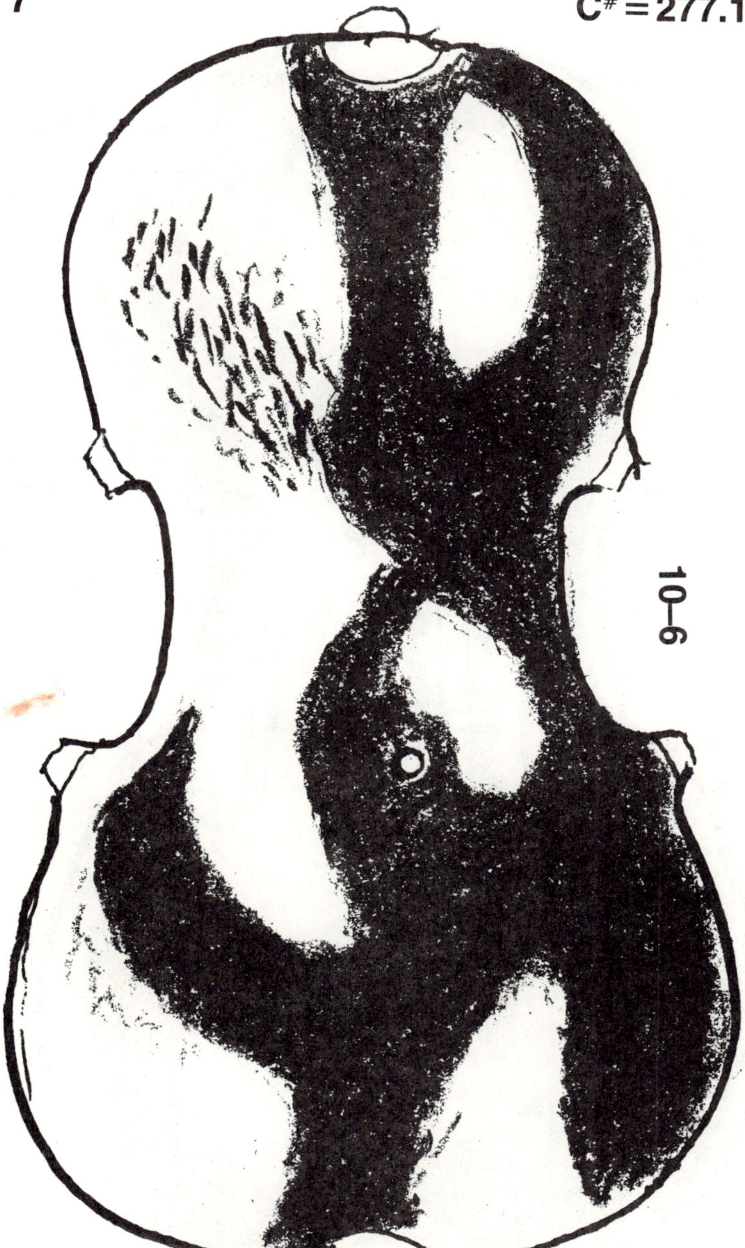

10–6

8

D = 293.66

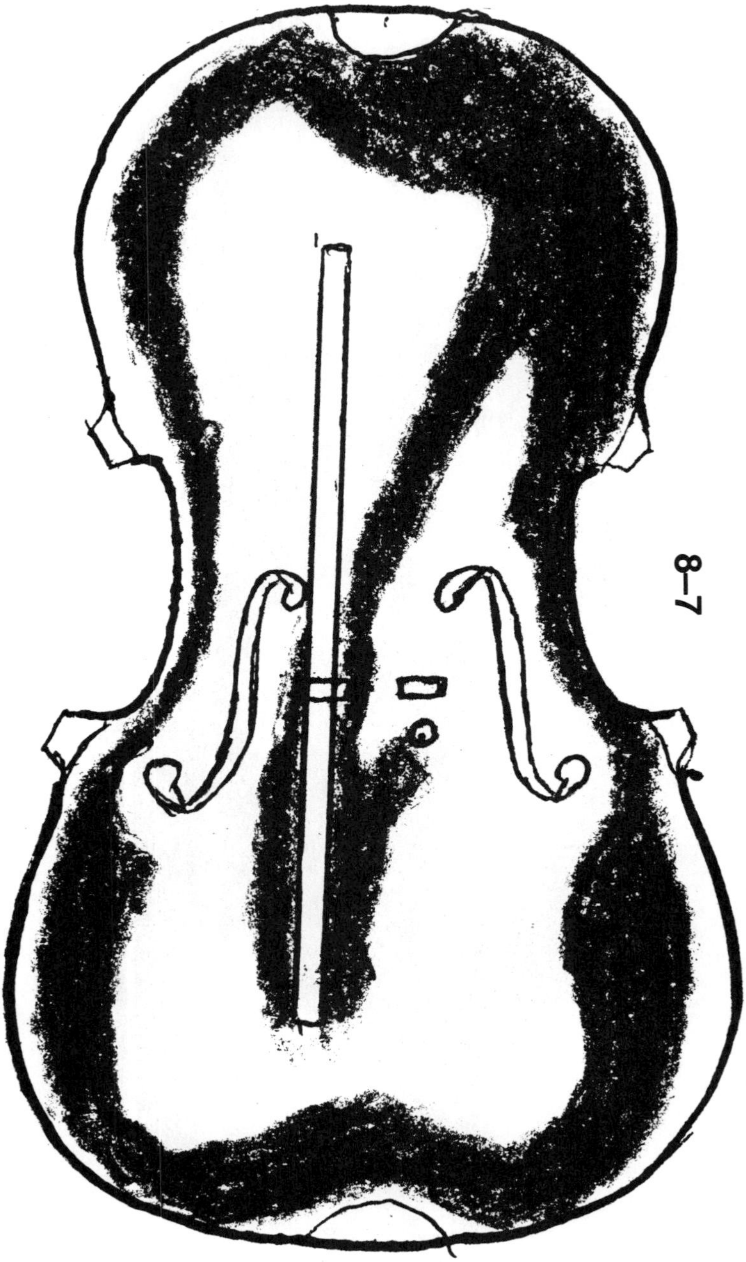

8–7

8 **D = 293.66**

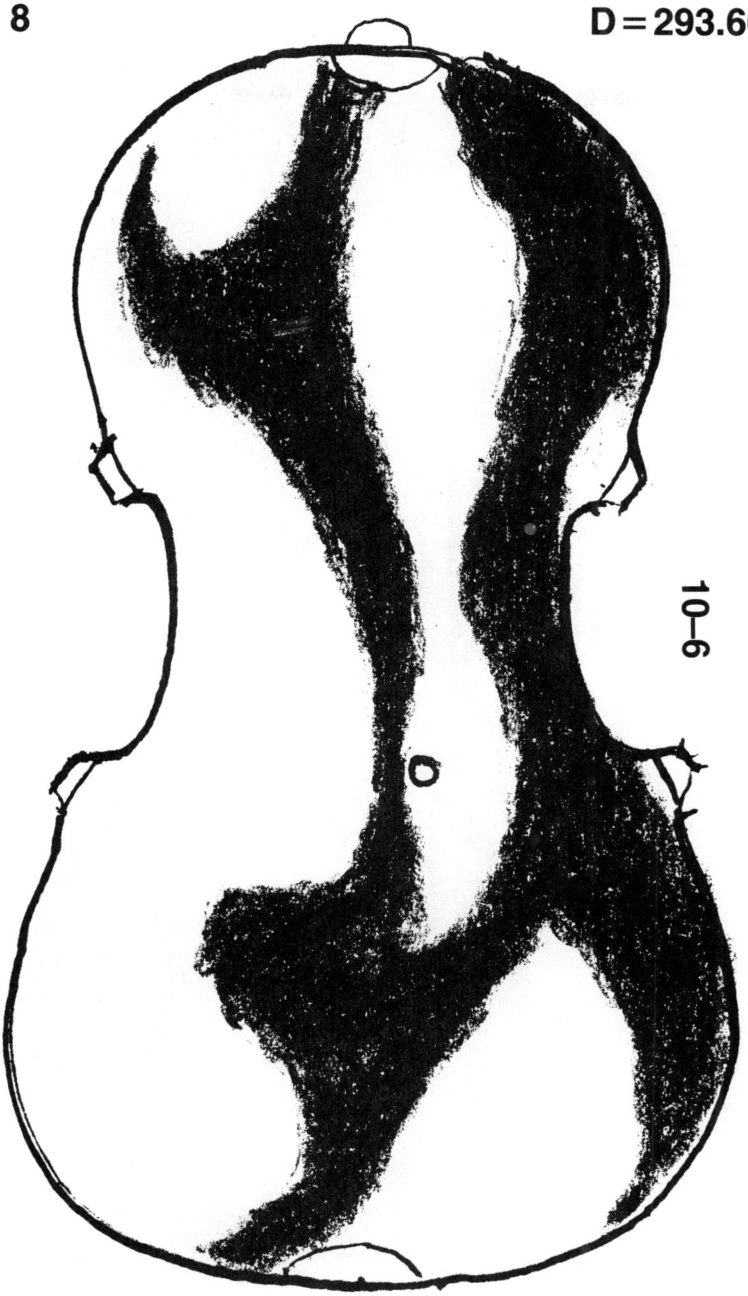

9

$D^{\#} = 311.13$

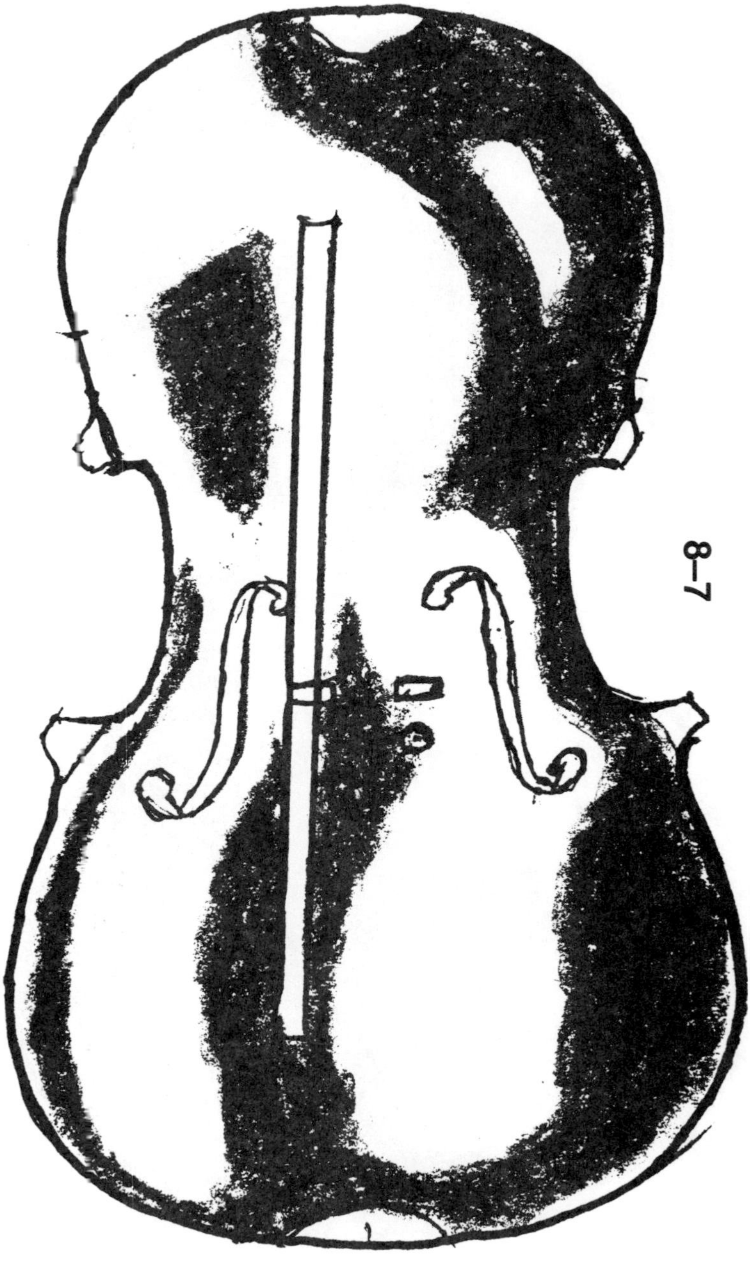

8–7

9 $D^{\#} = 311.13$

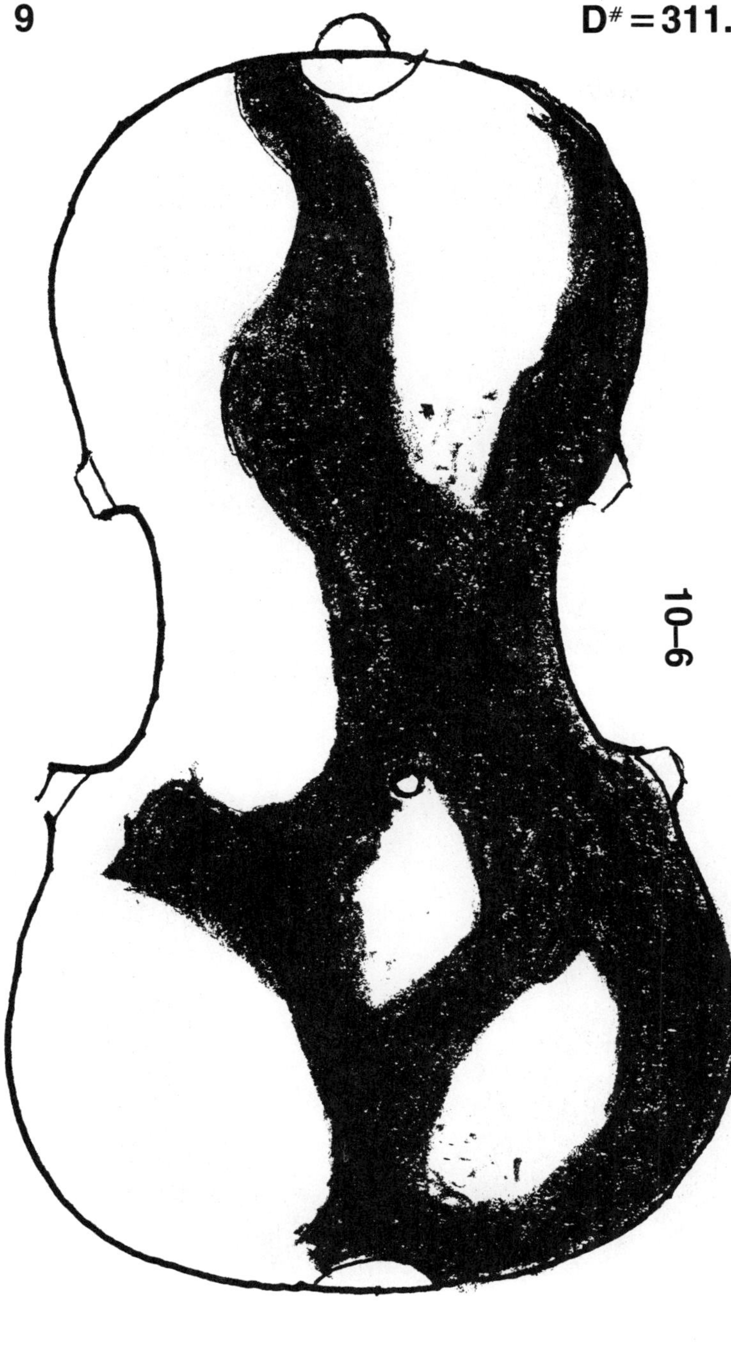

10–6

10 E = 329.63

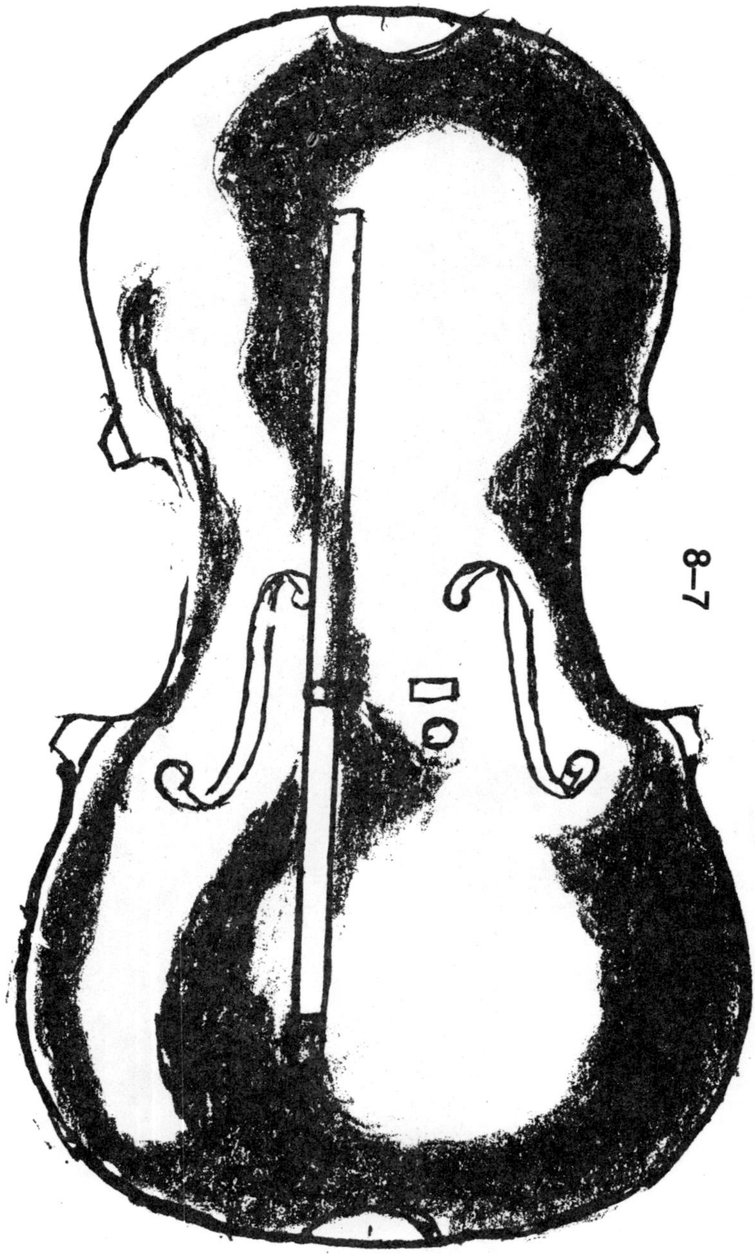

10

E = 329.63

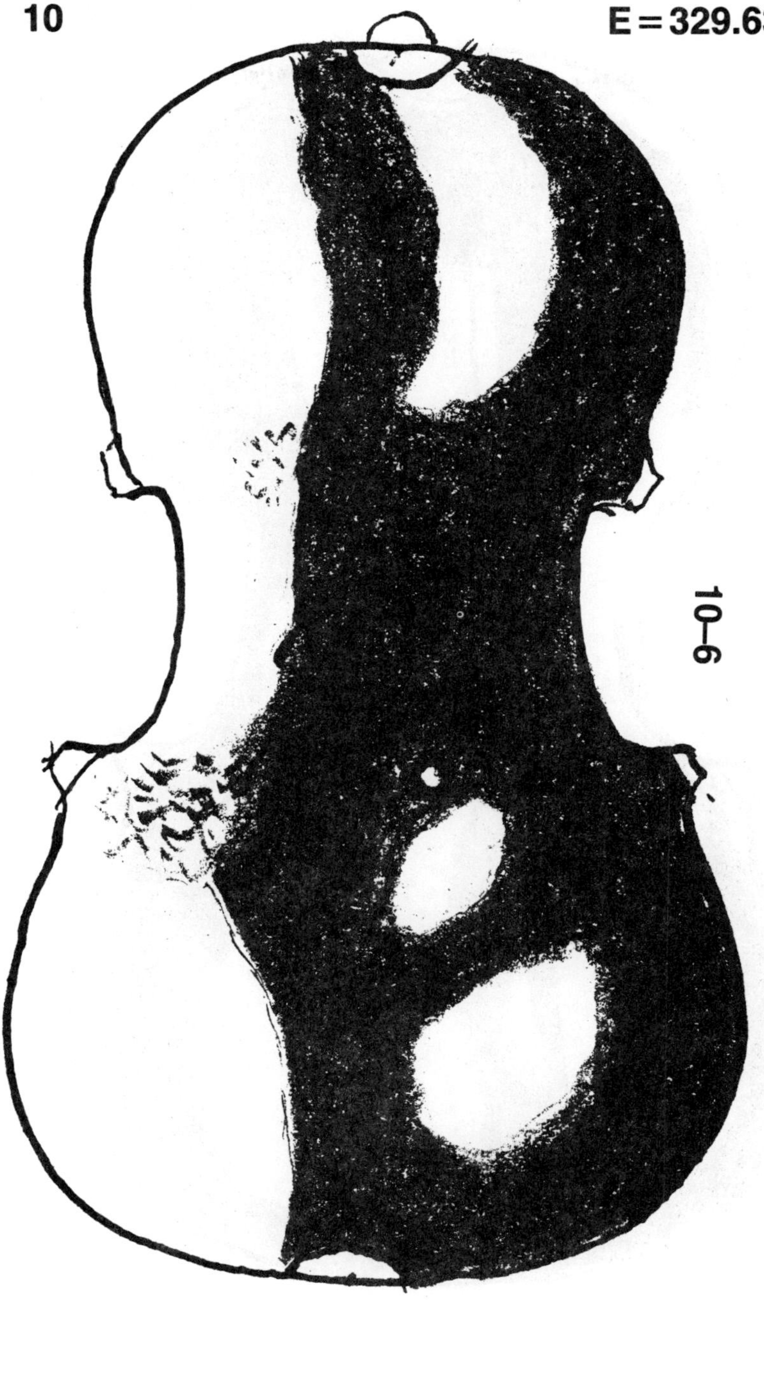

10–6

11

F = 349.23

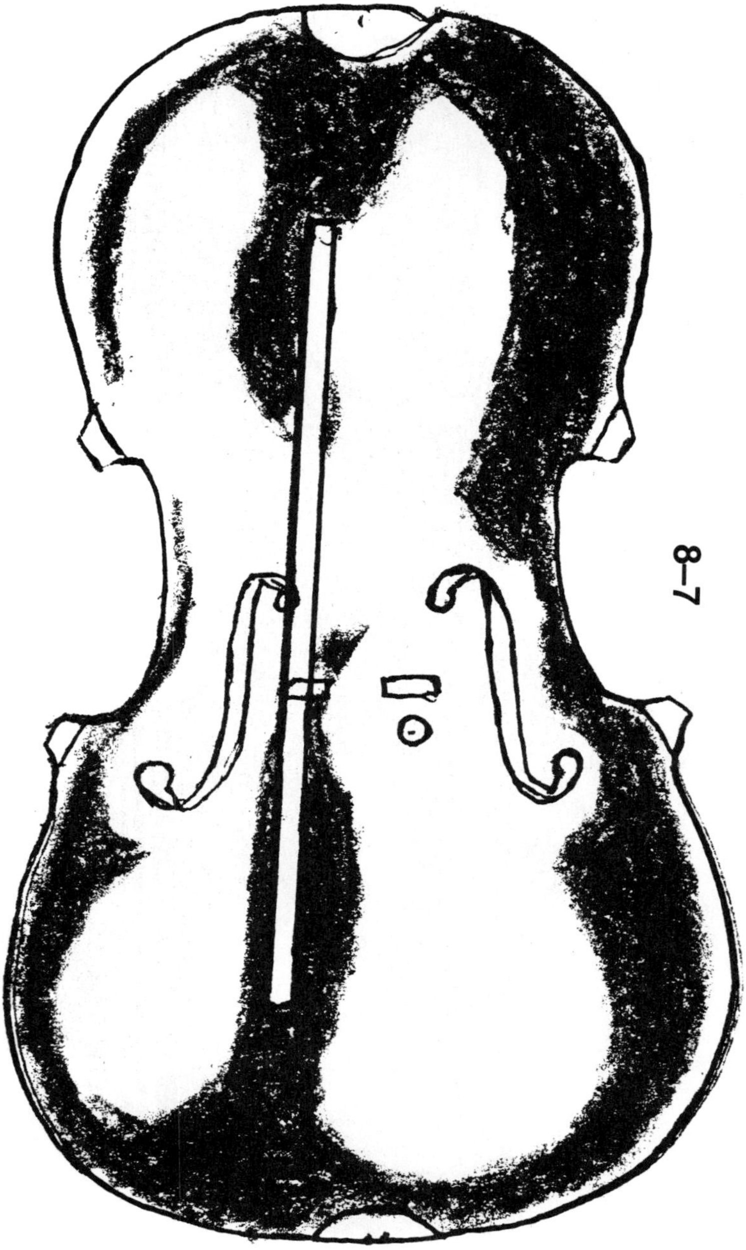

8–7

11

F = 349.23

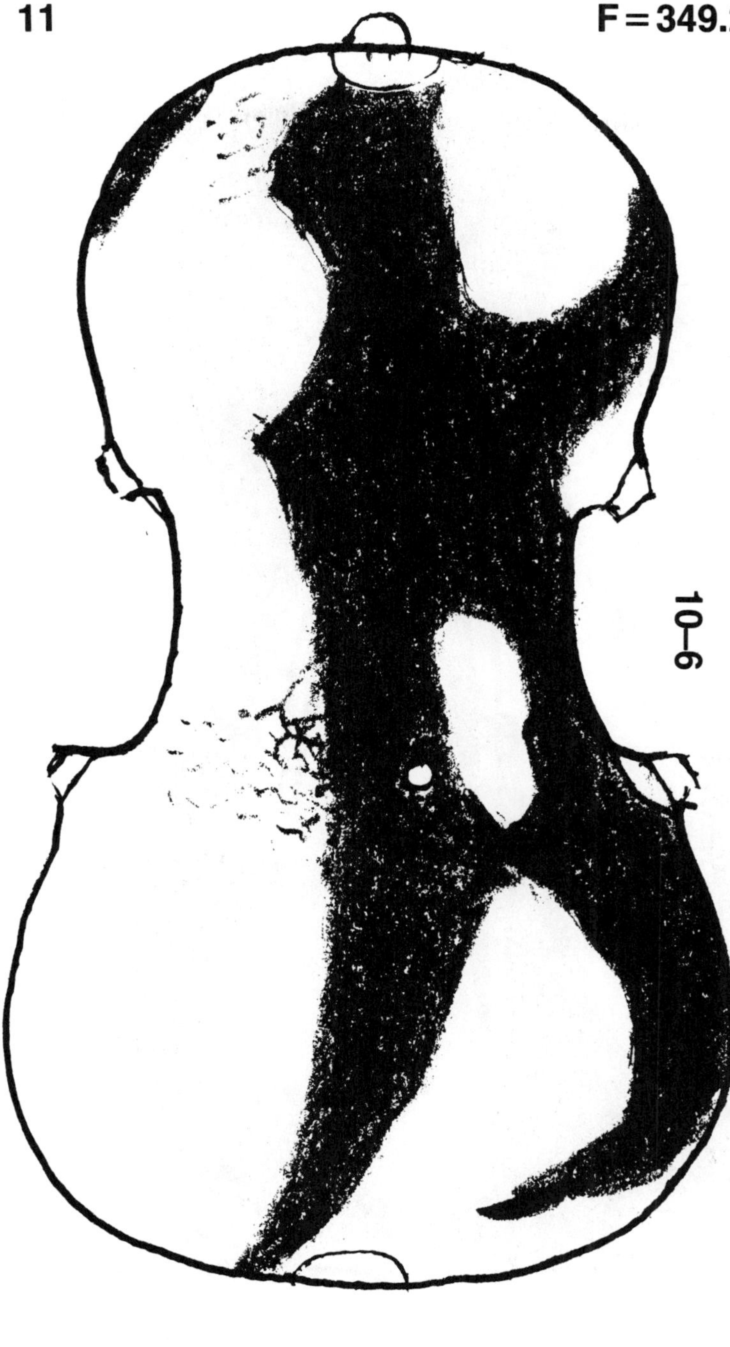

10–6

12

$F^{\#} = 370$

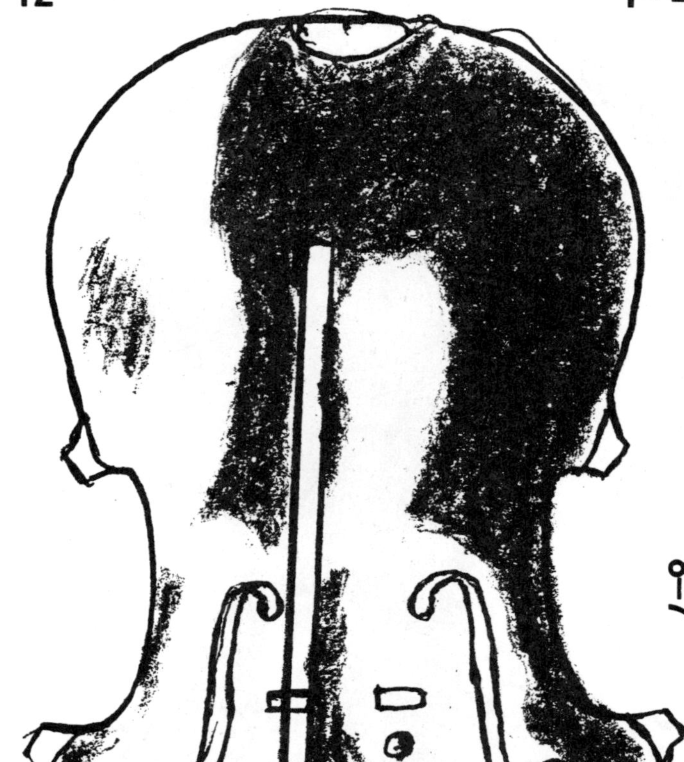

8–7

12

F# = 370

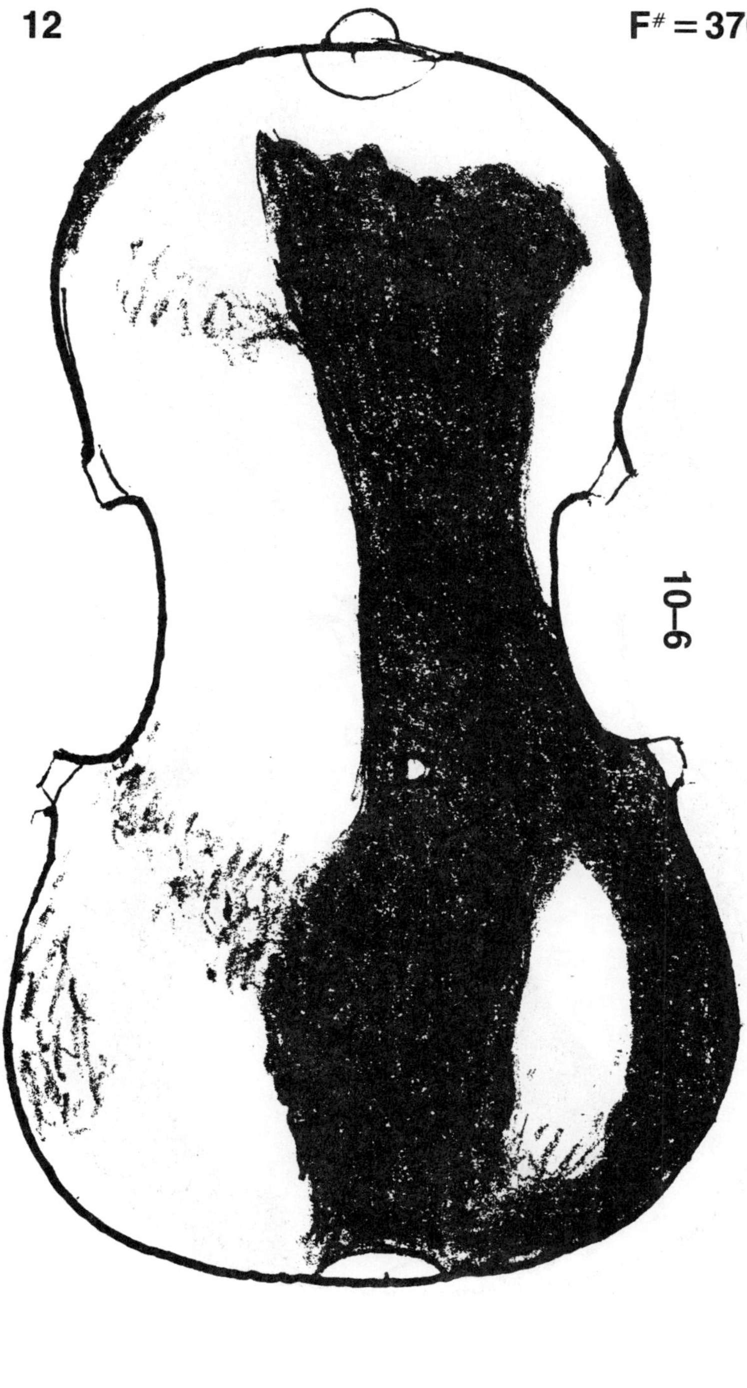

10–6

13

G = 392

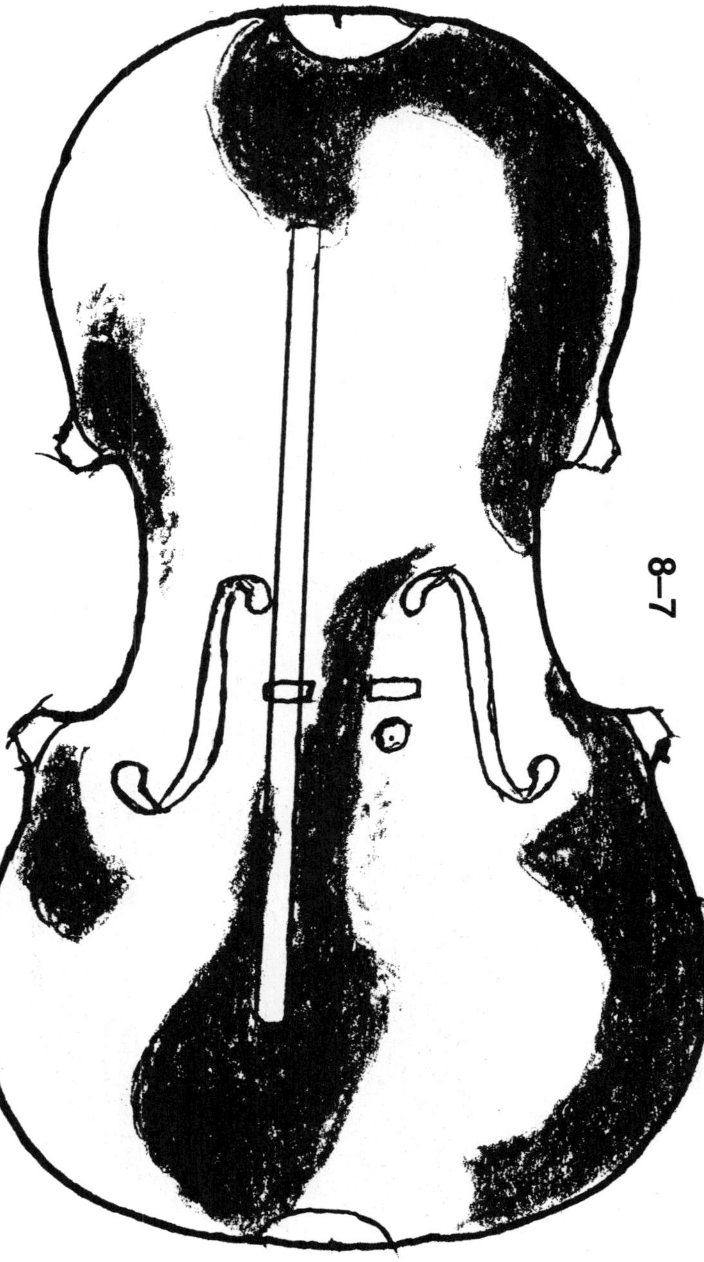

8–7

13 G = 392

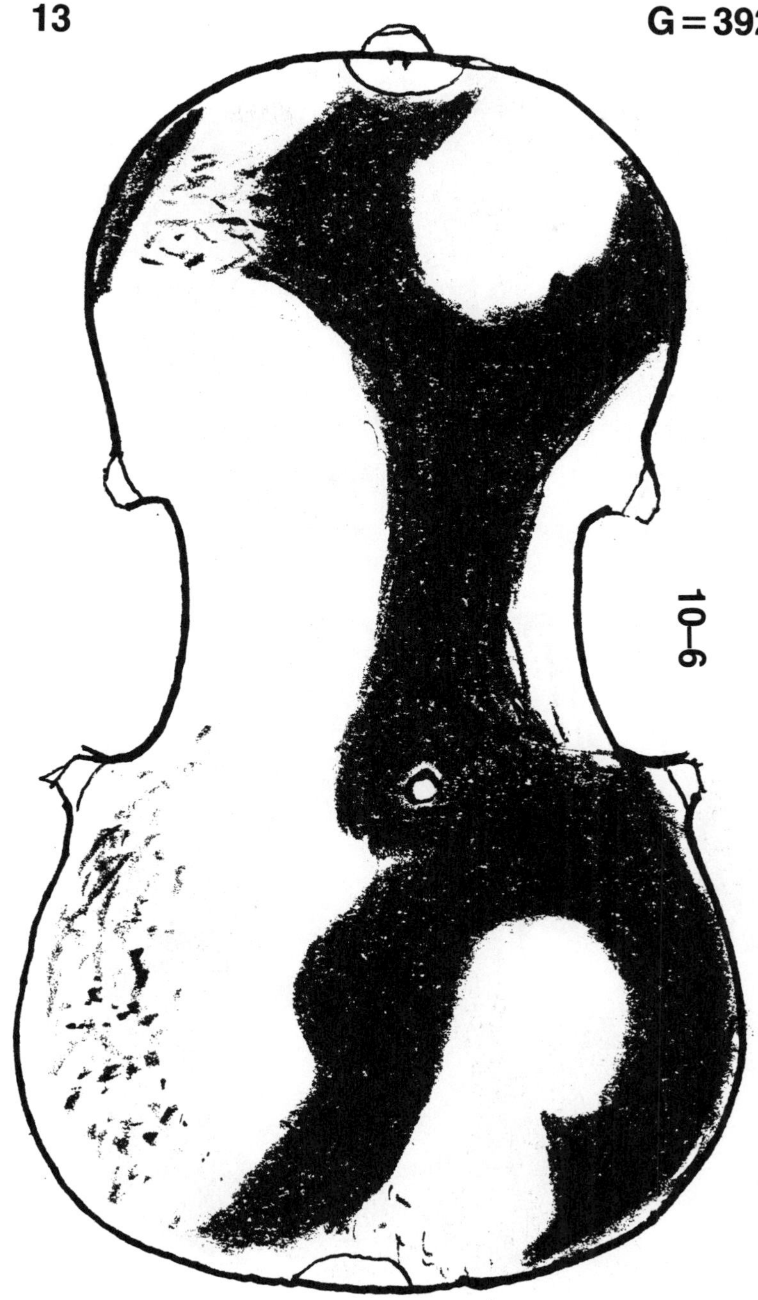

14 $G^{\#} = 415.3$

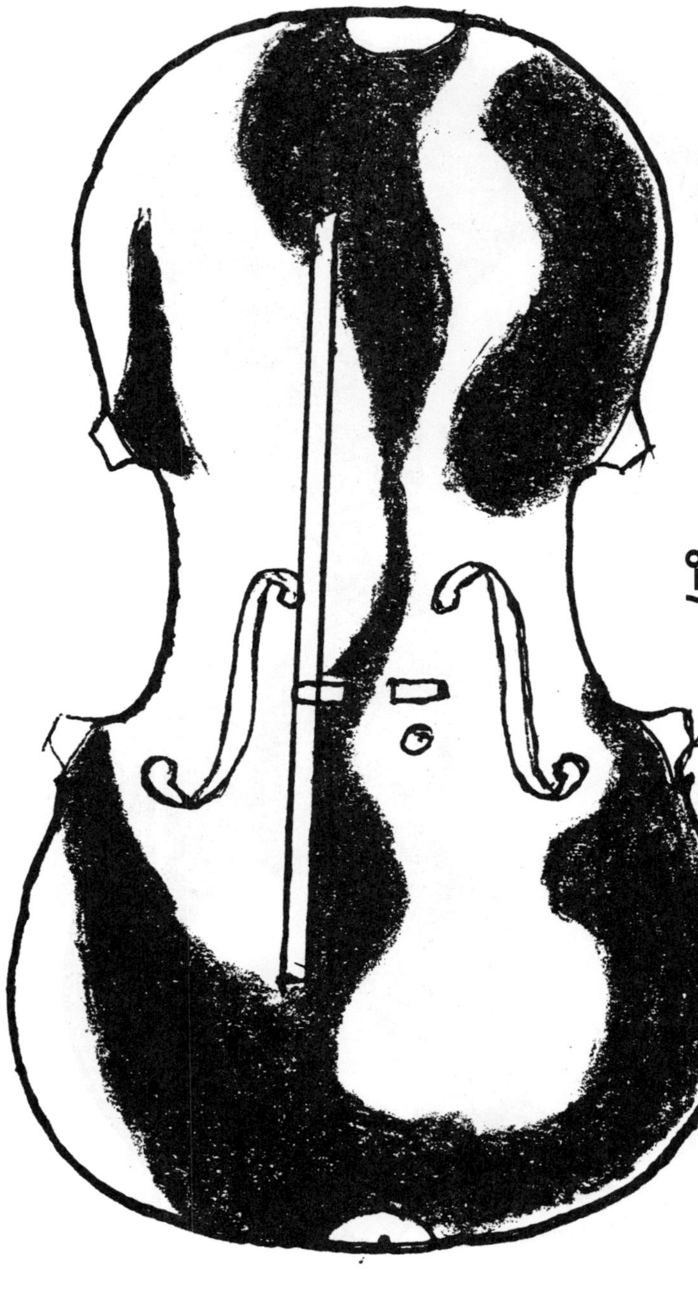

8–7

14

$G^{\#} = 415.3$

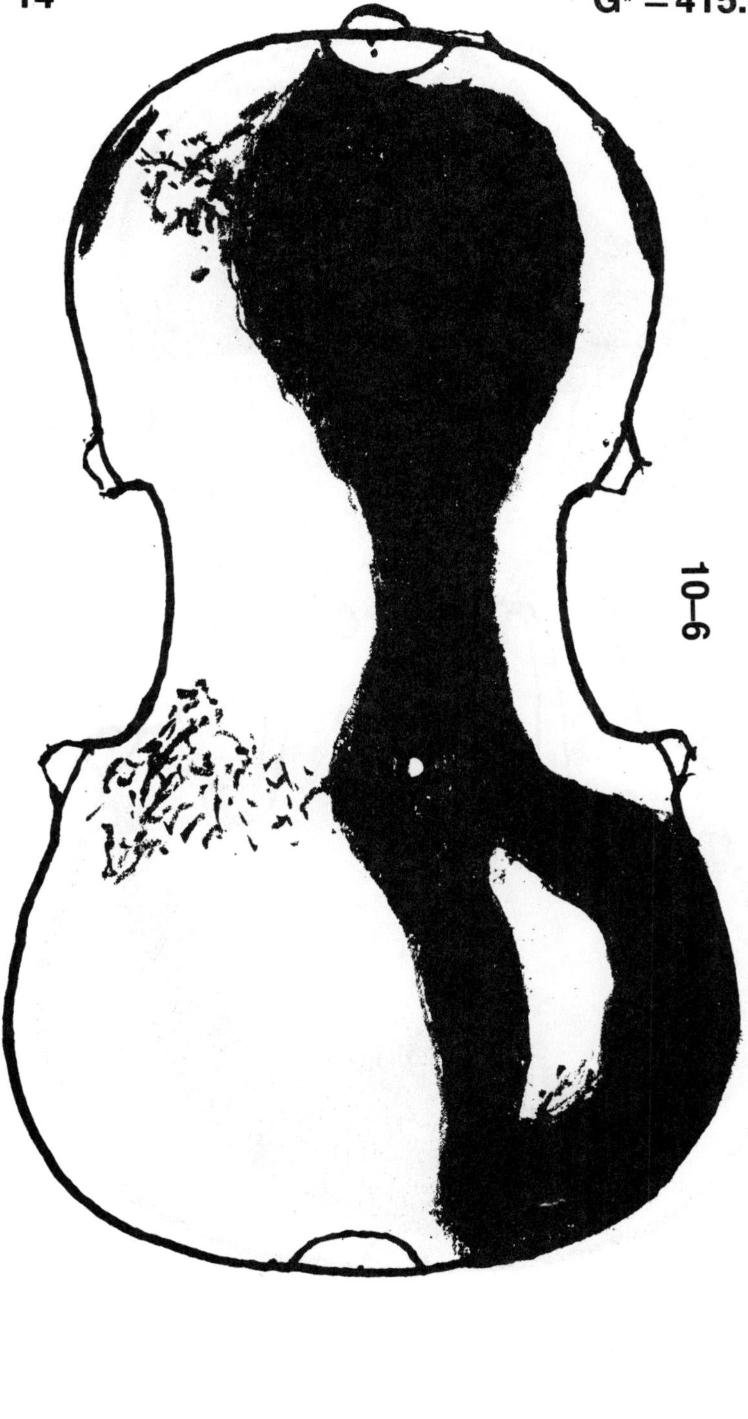

10–6

15

A = 440

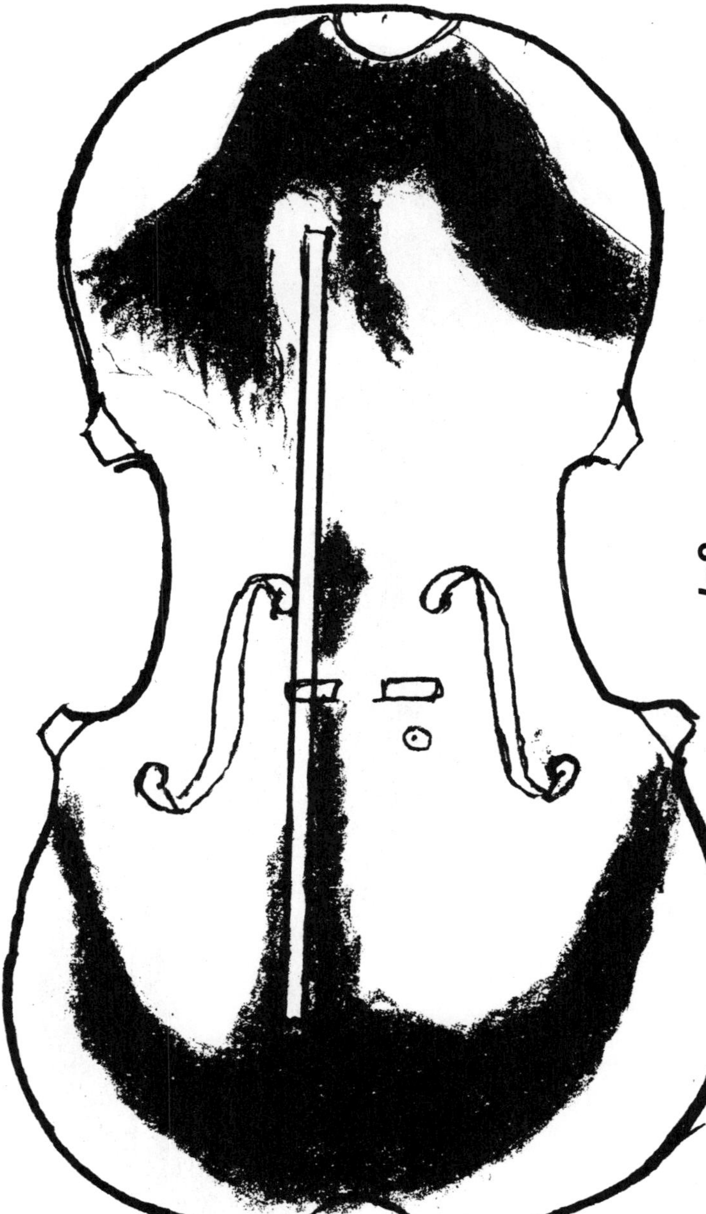

8–7

15

A = 440

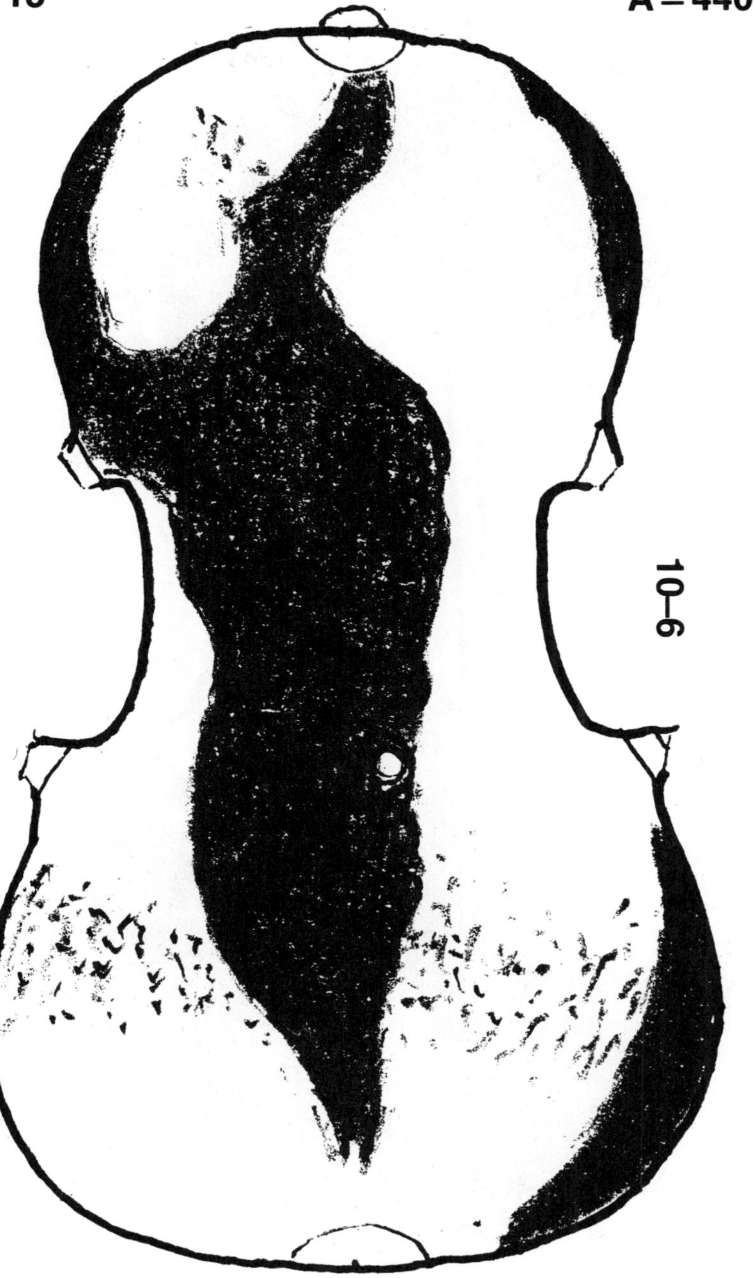

10–6

16 A# = 466.16

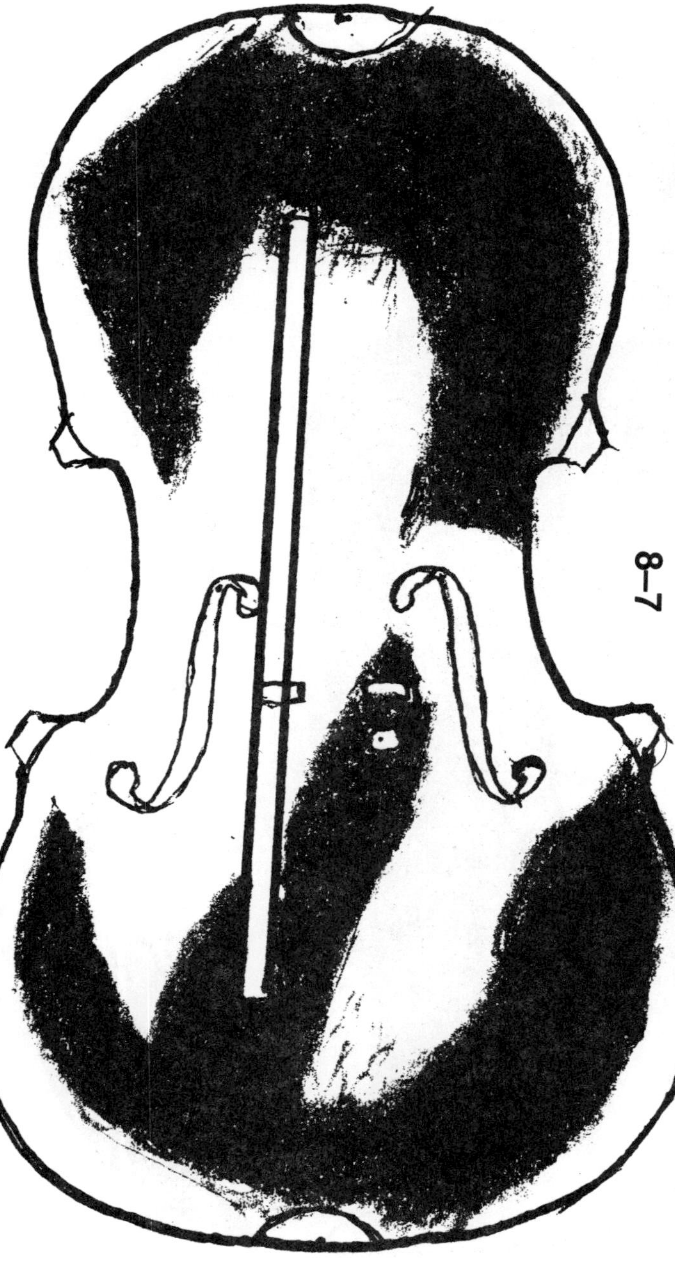

8–7

15

A = 440

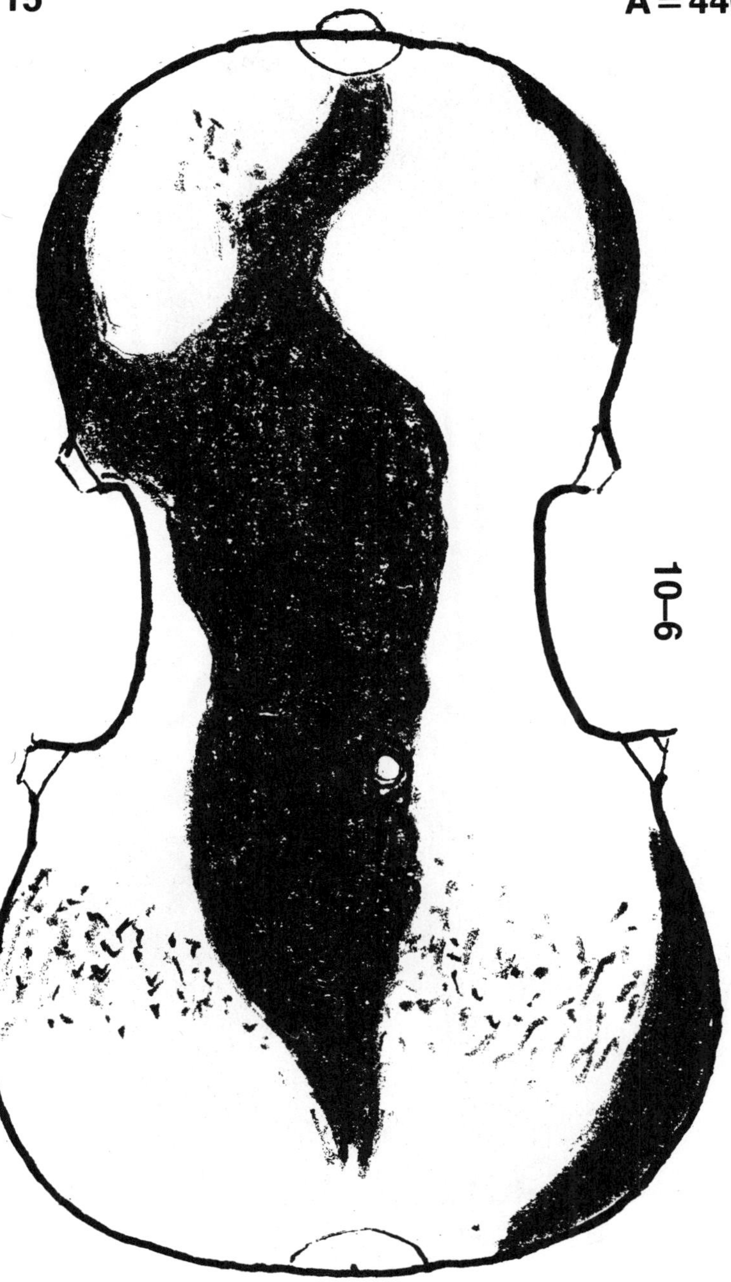

10–6

16 A$^{\#}$ = 466.16

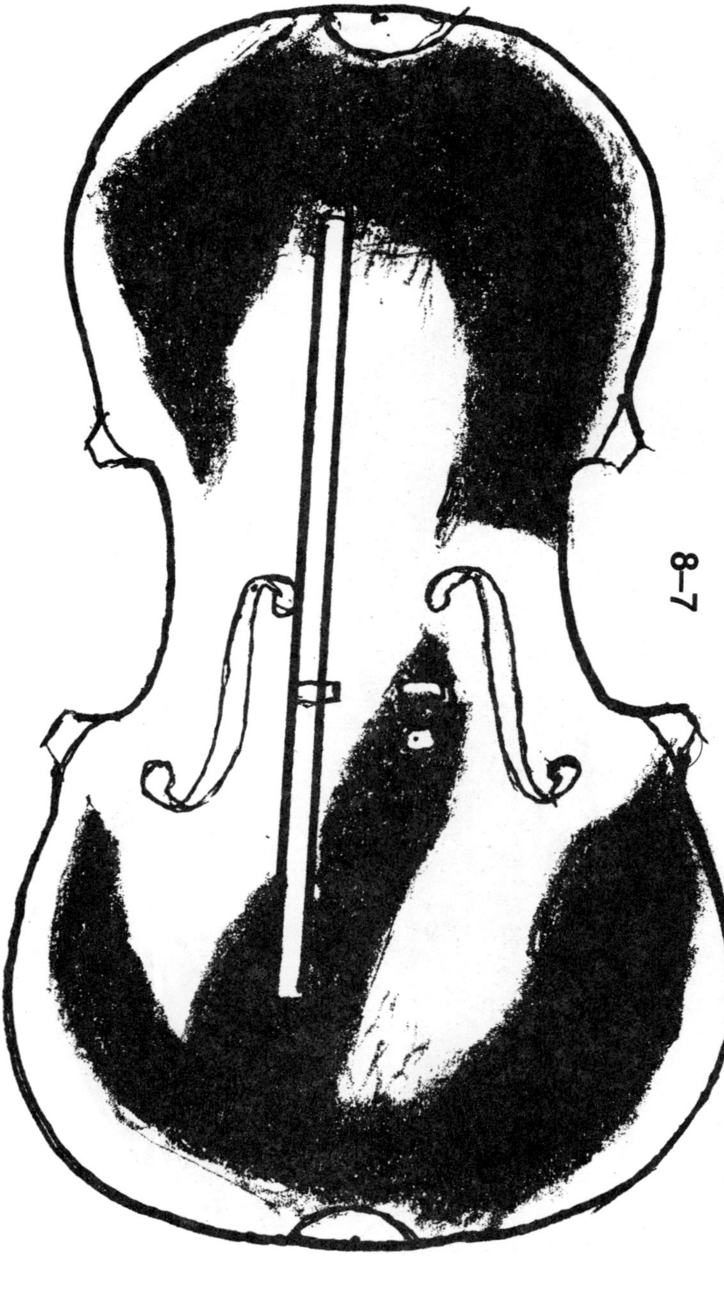

8–7

16

$A^{\#} = 466.16$

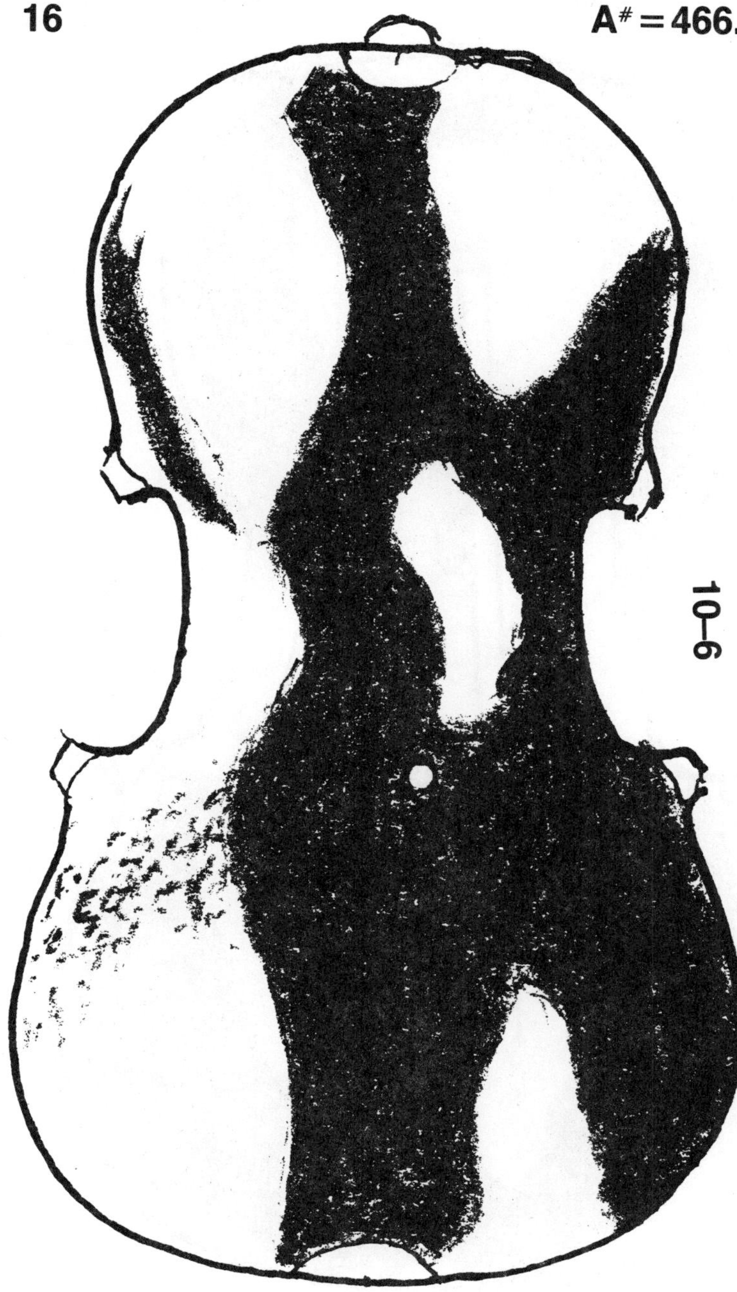

10–6

17 B = 493.88

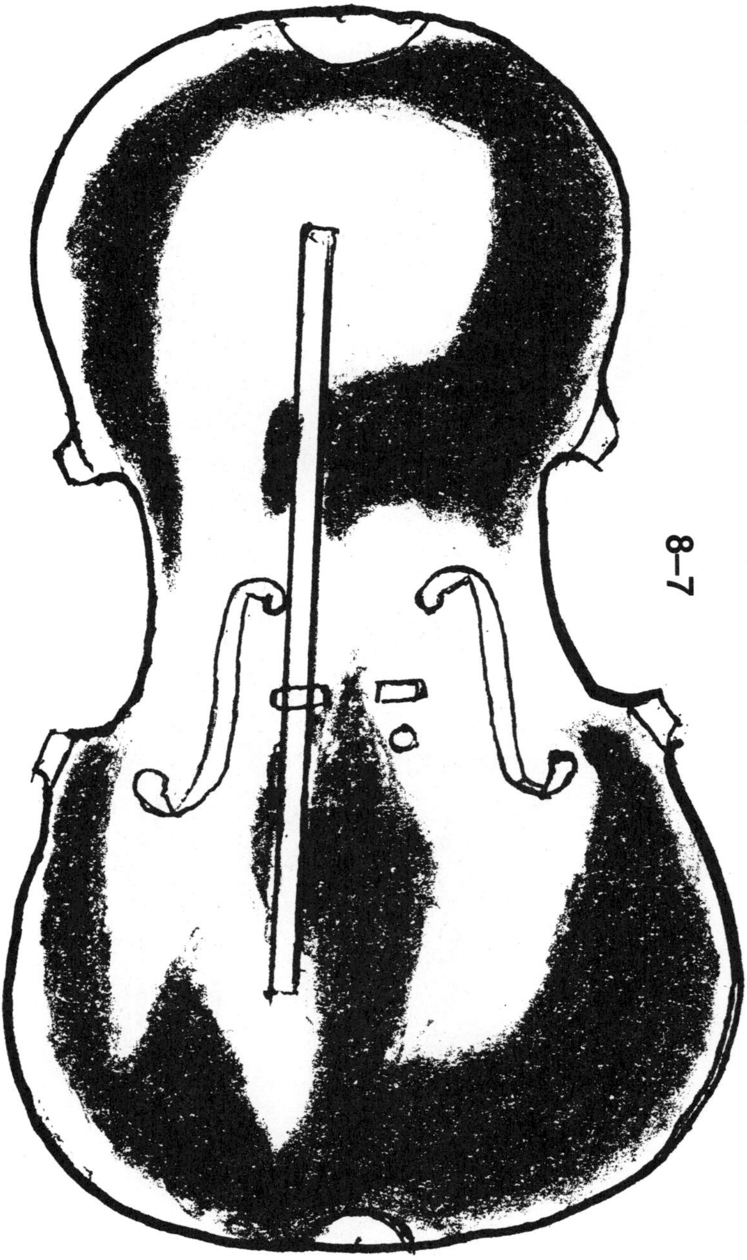

8–7

17

B = 493.88

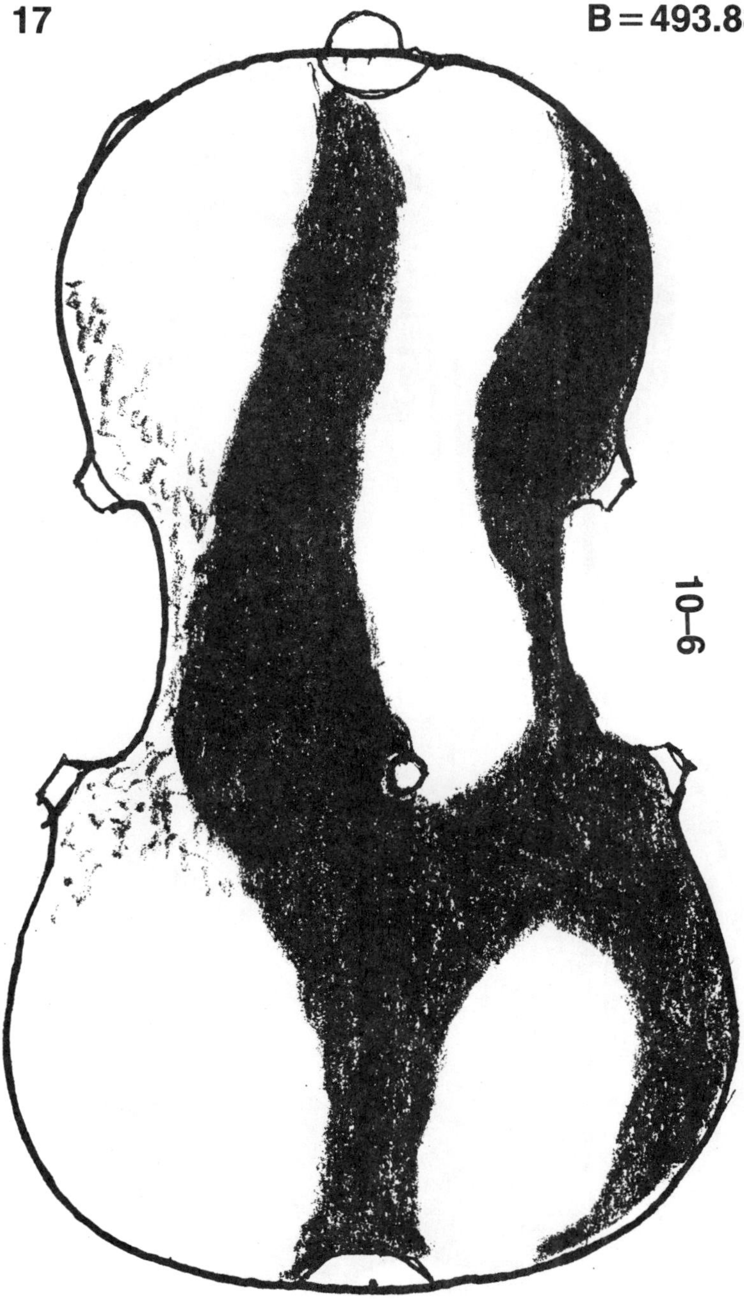

10–6

18 C = 523.25

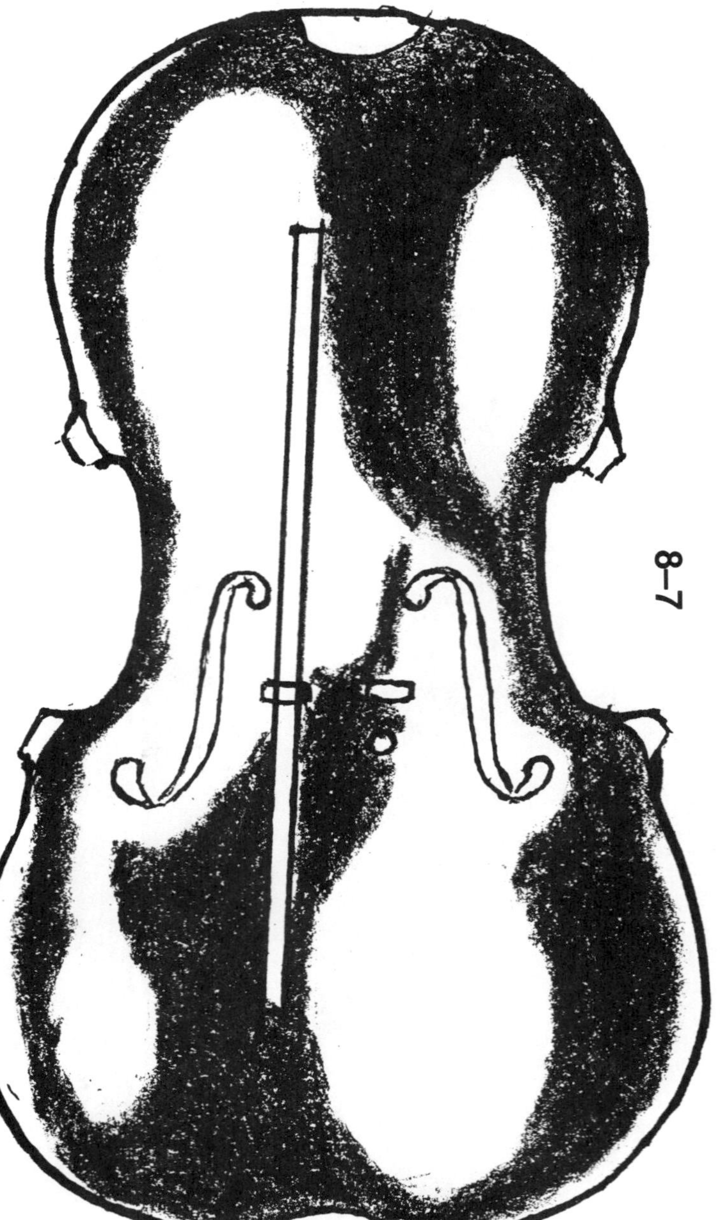

8–7

18

C = 523.25

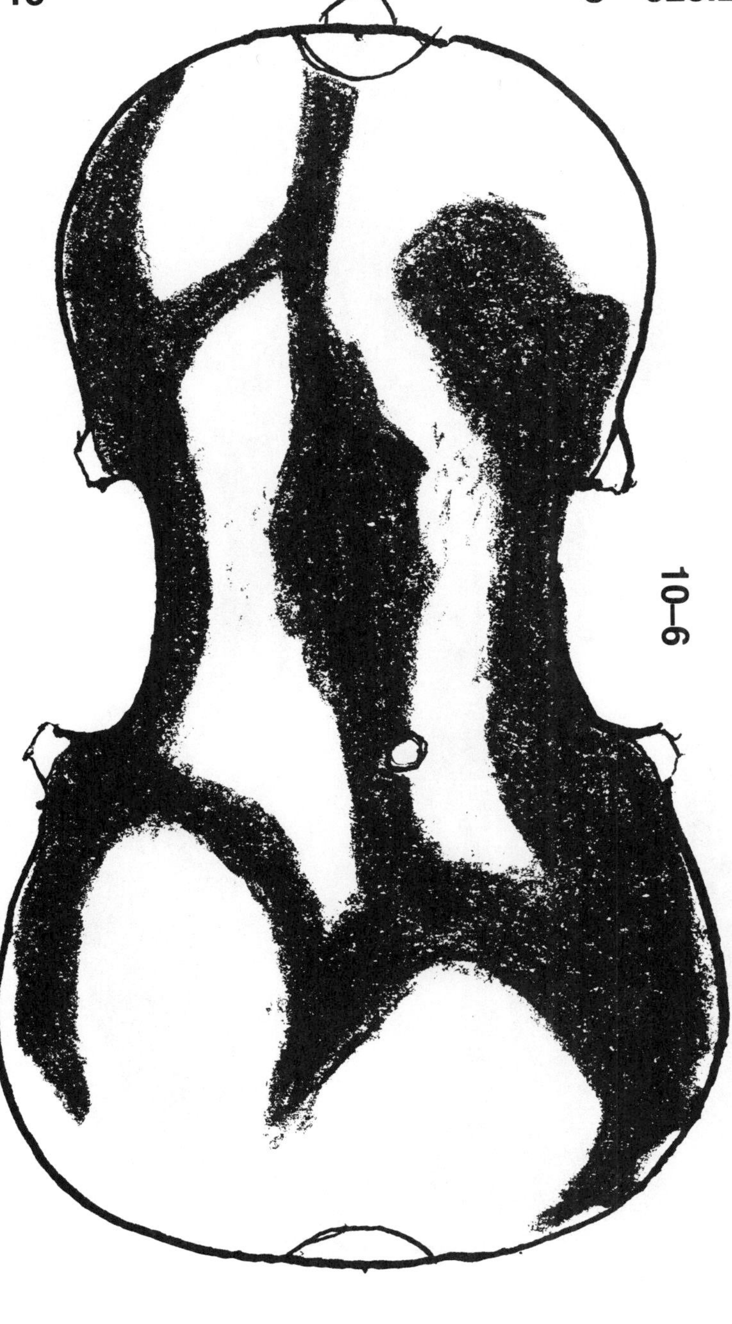

10–6

19

$C^{\#} = 554.37$

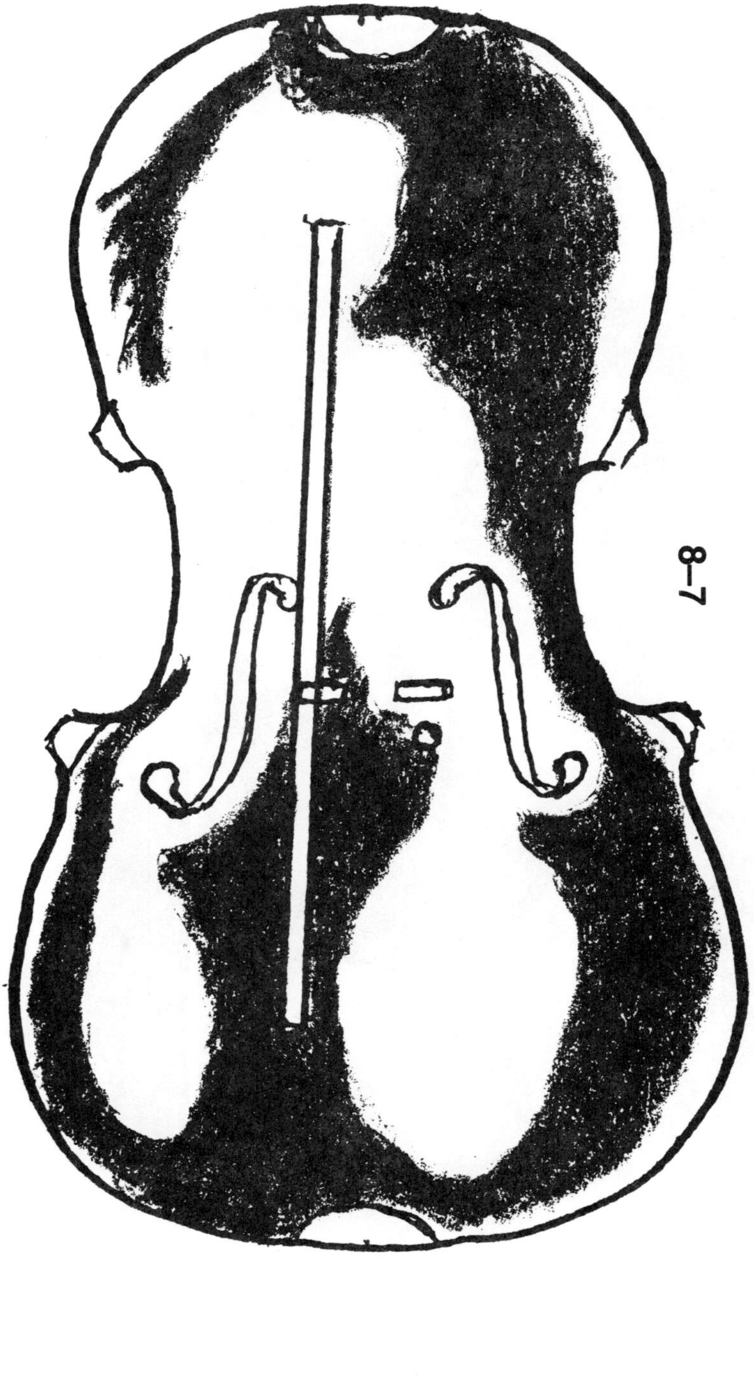

8–7

19

$C^{\#} = 554.37$

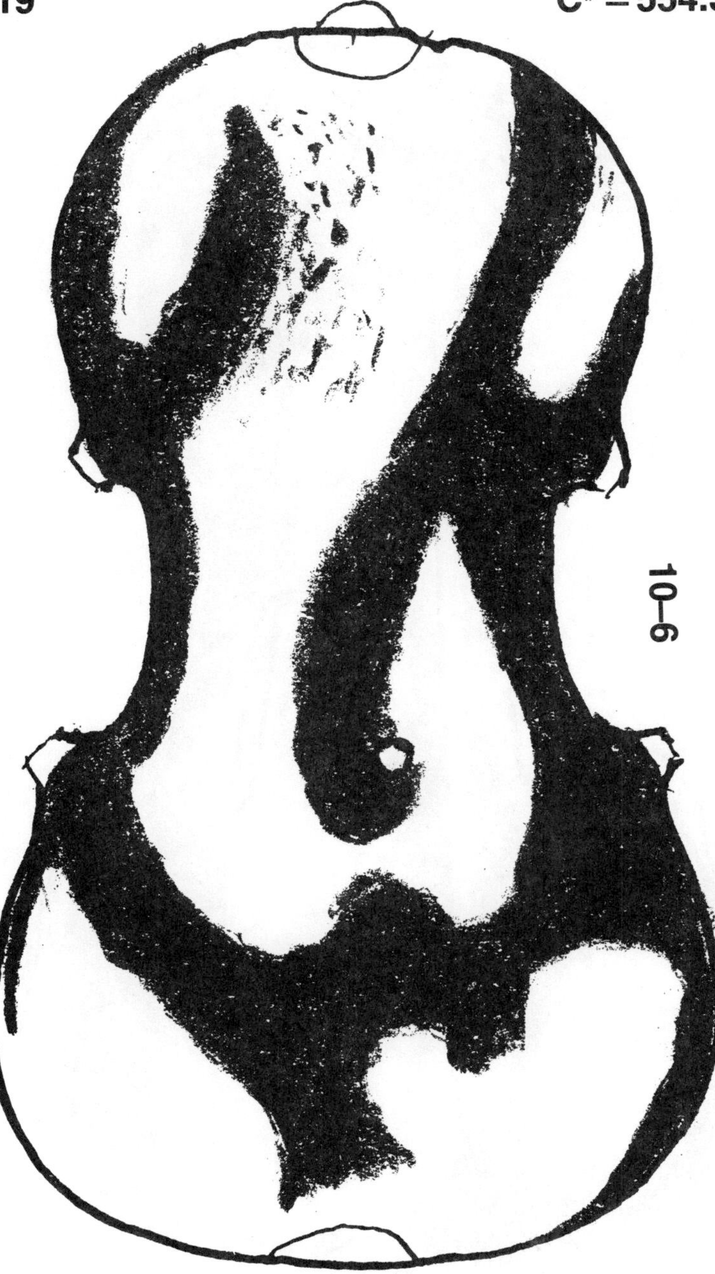

10–6

20 D = 587.33

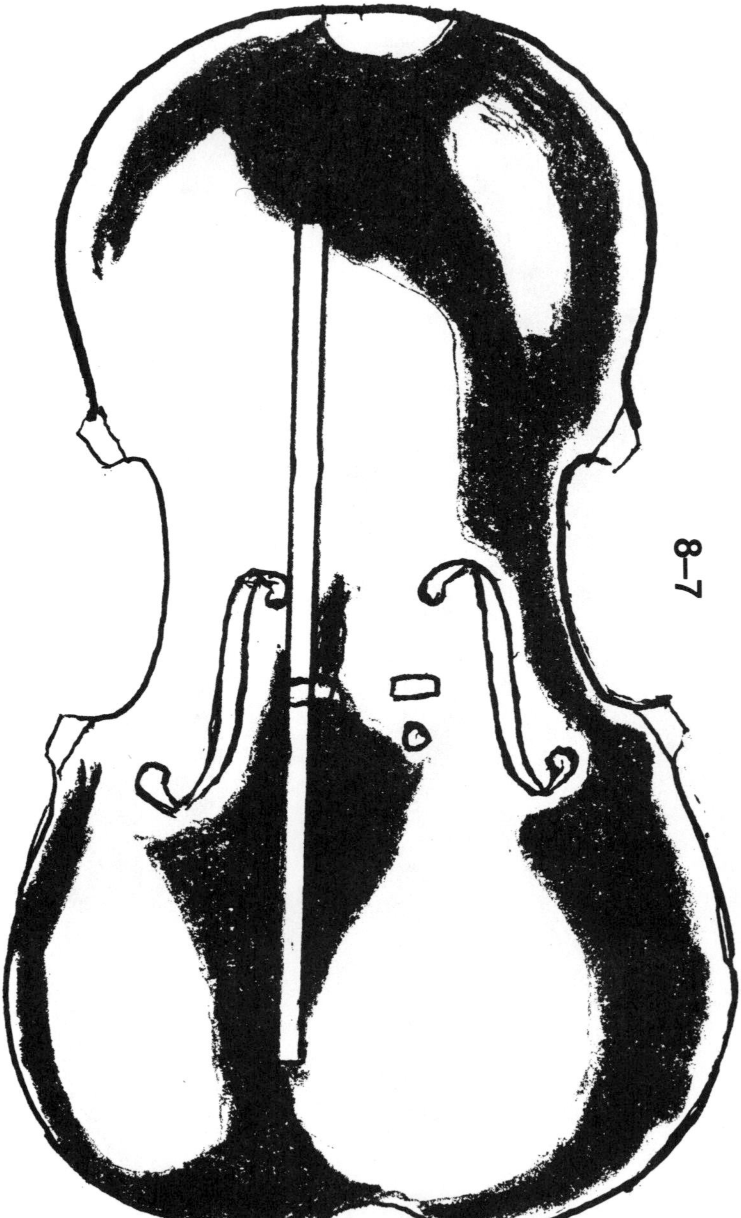

20

D = 587.33

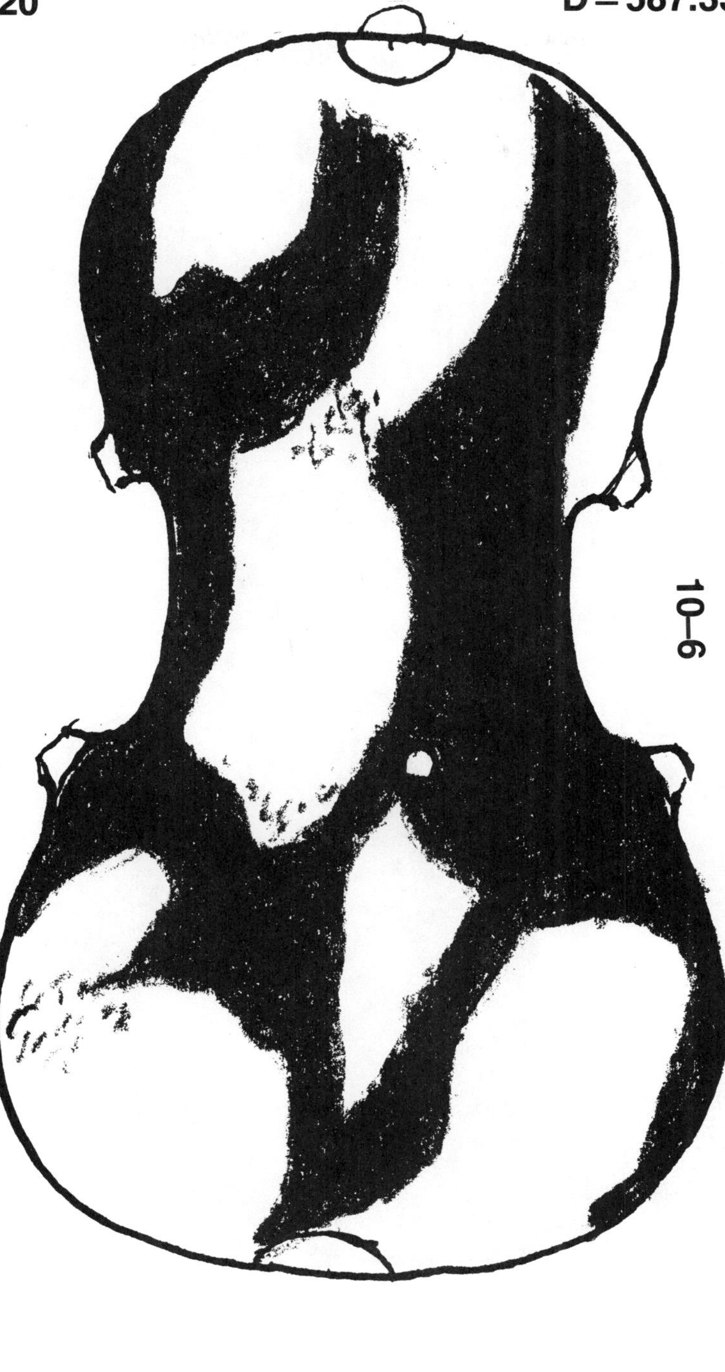

10–6

21

$D^{\#} = 622.25$

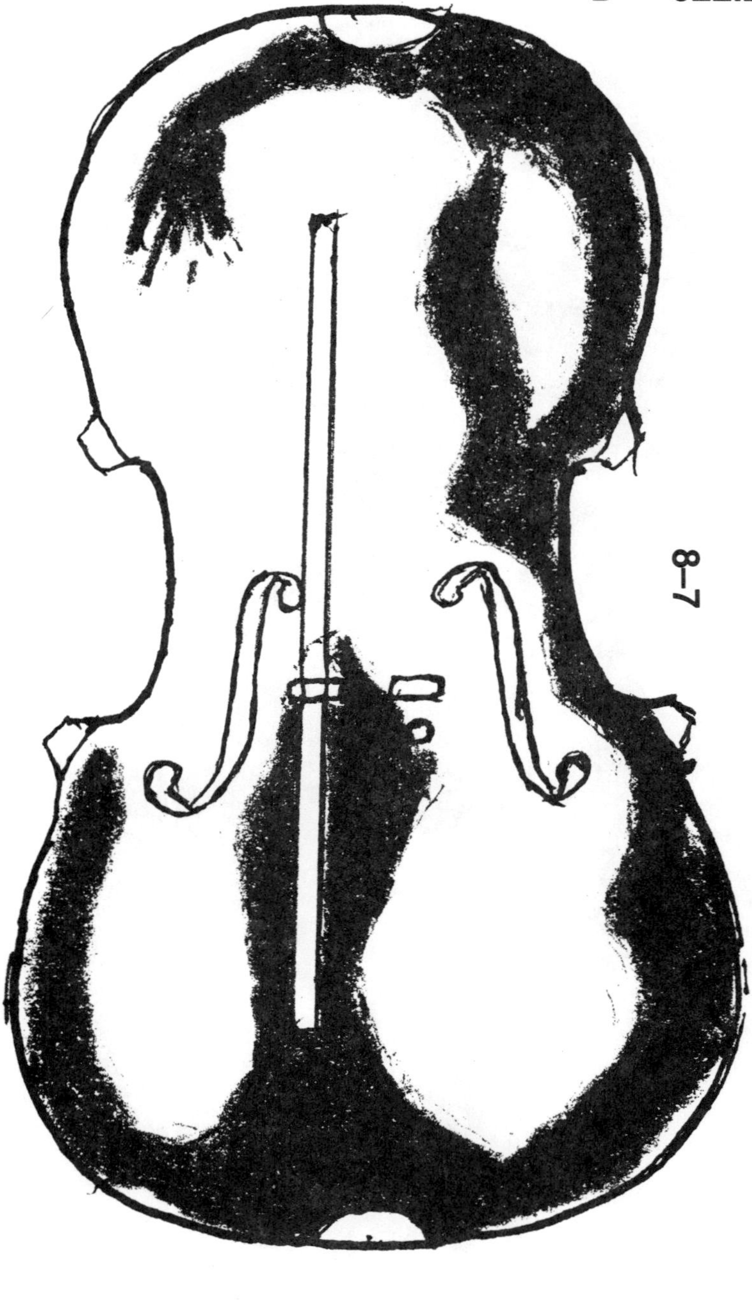

8–7

21

$D^{\#} = 622.25$

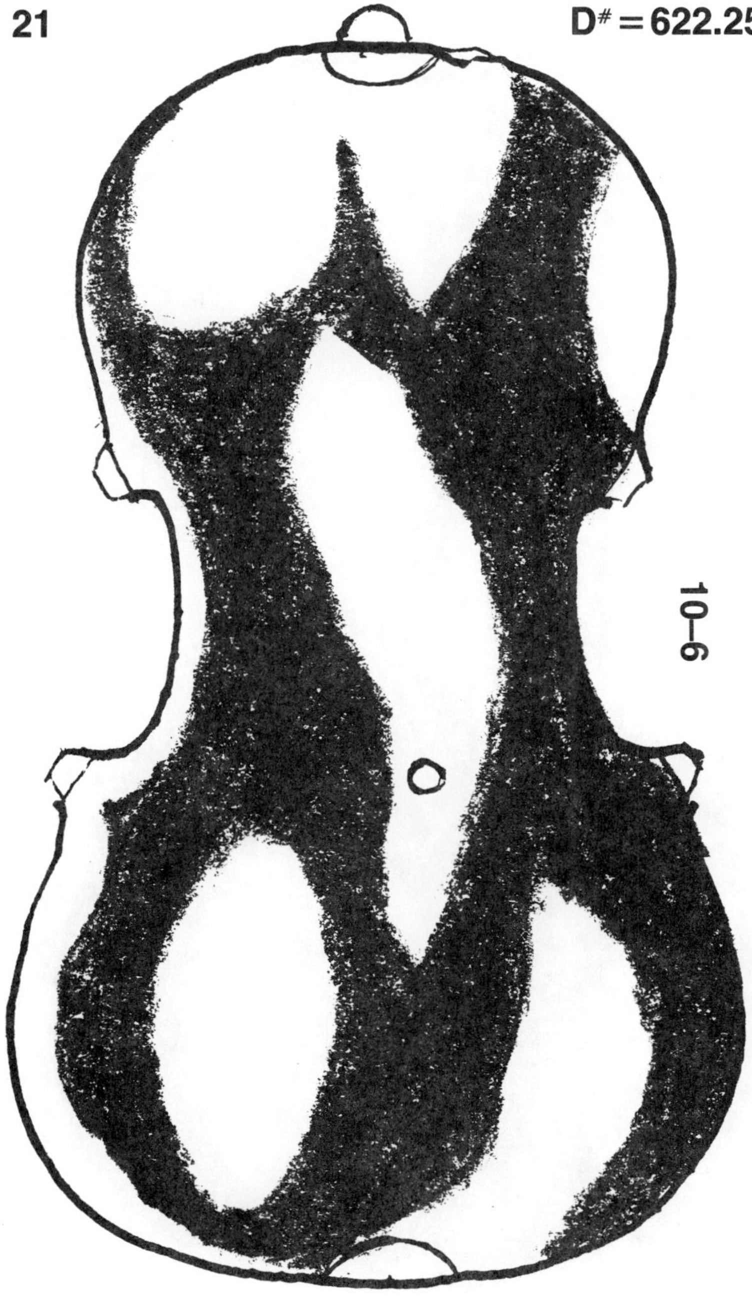

22

E = 659.26

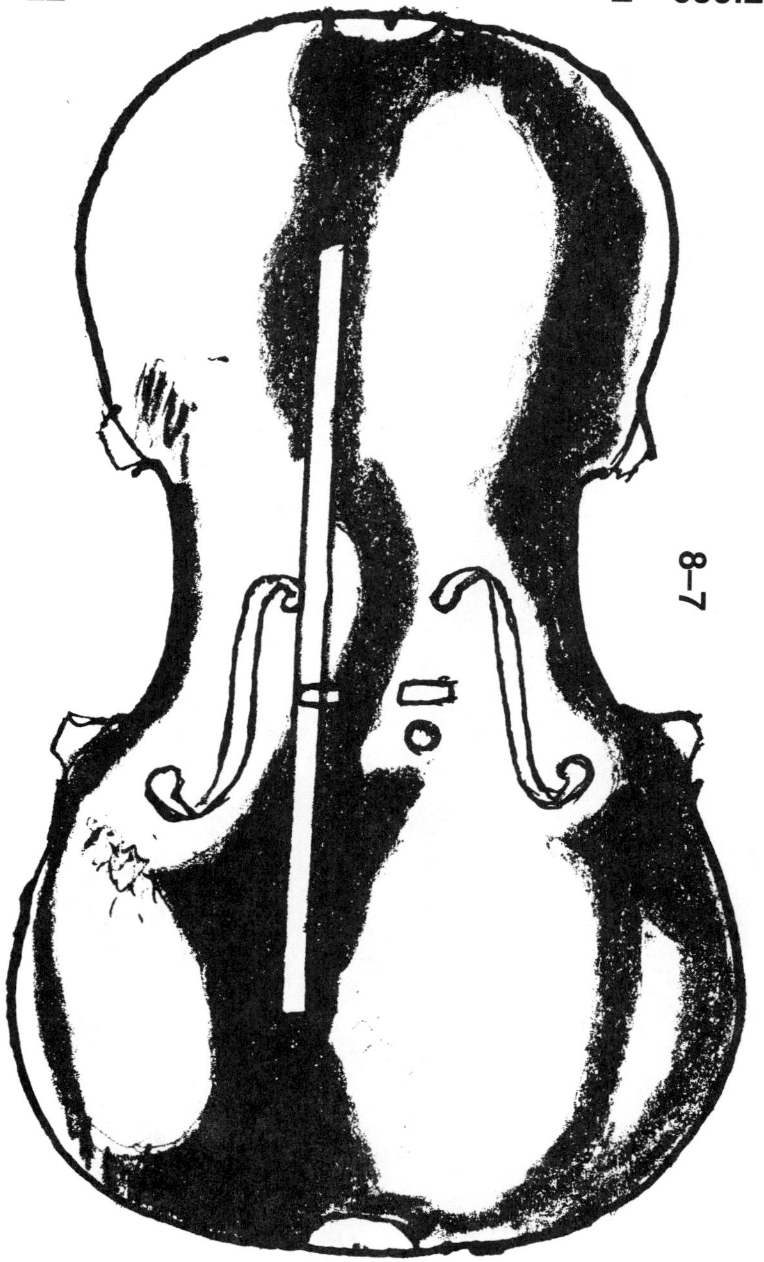

8–7

22 E = 659.26

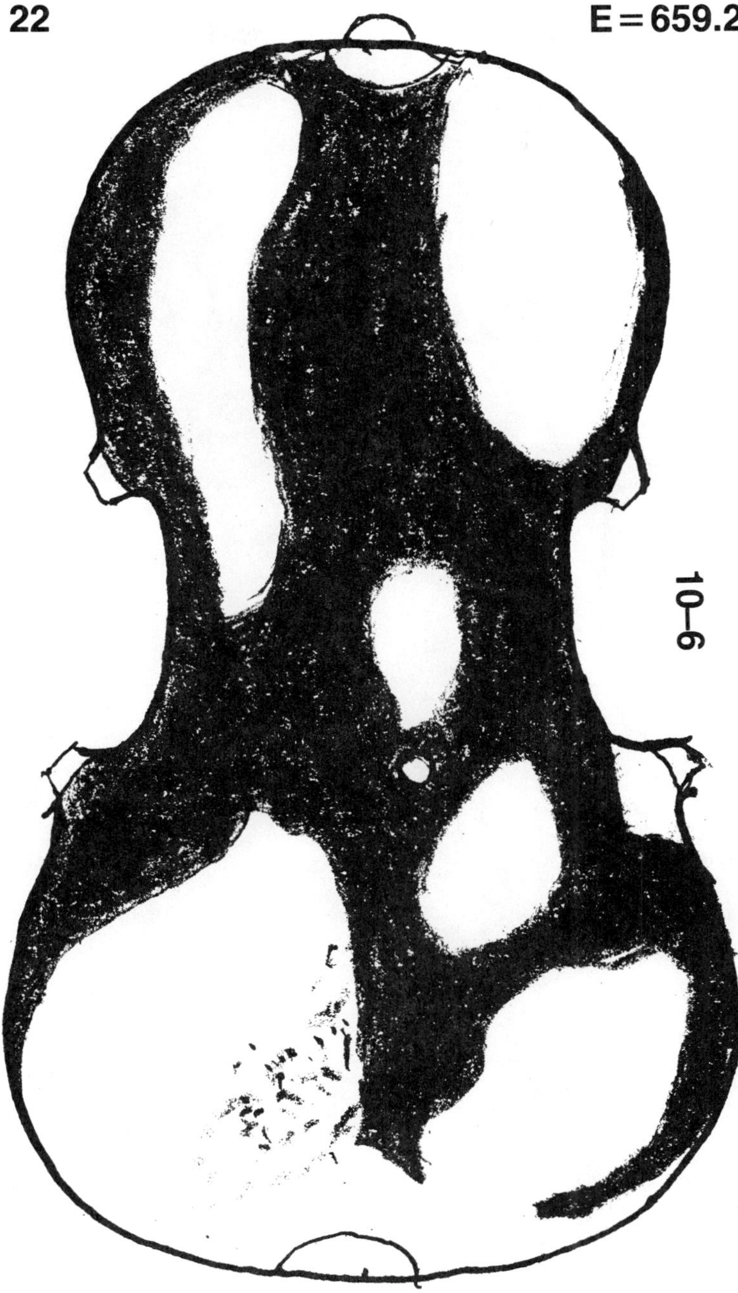

10–6

23 F = 698.46

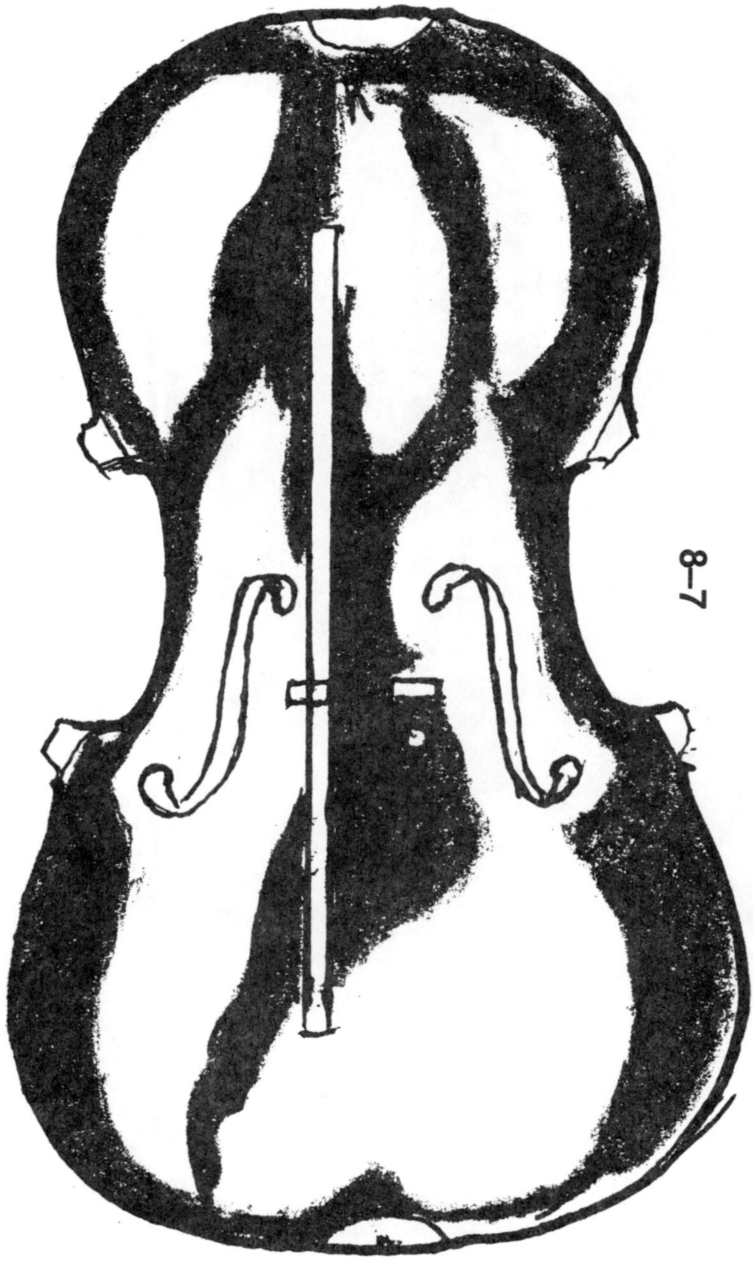

8–7

23

F = 698.46

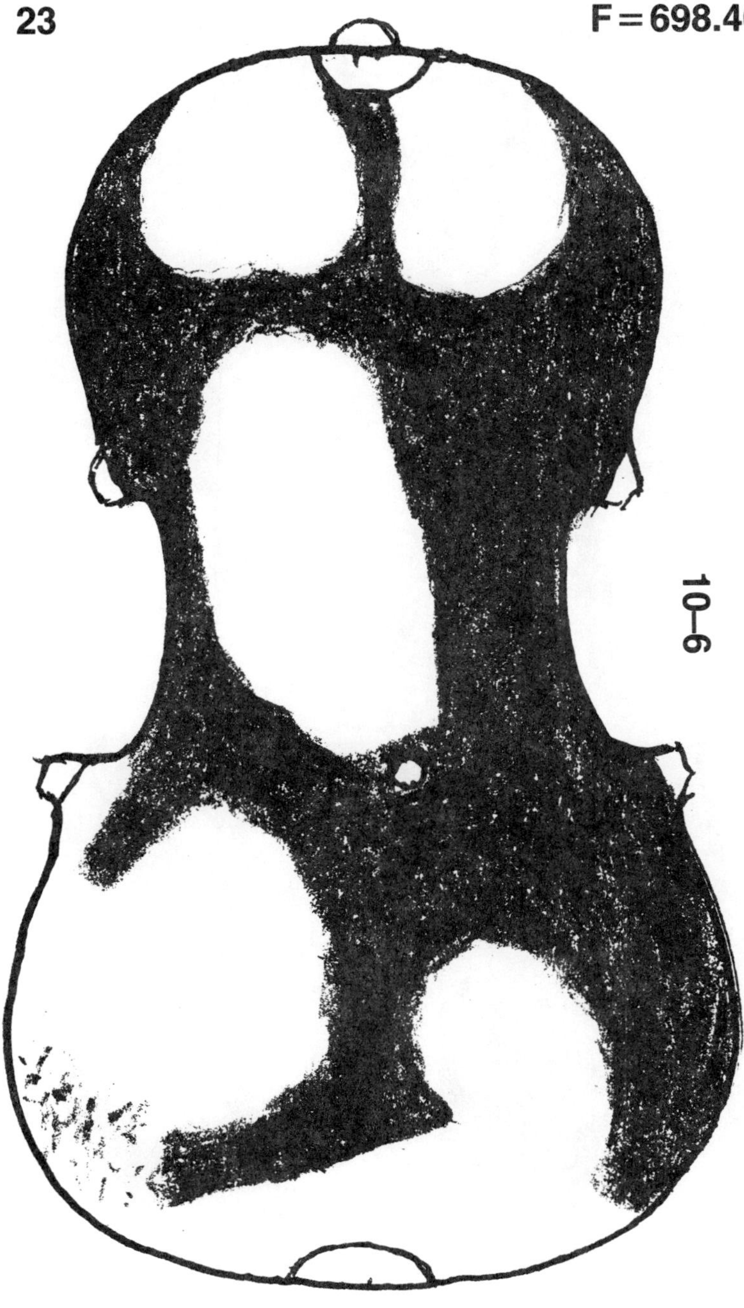

10–6

24

$F^{\#} = 740$

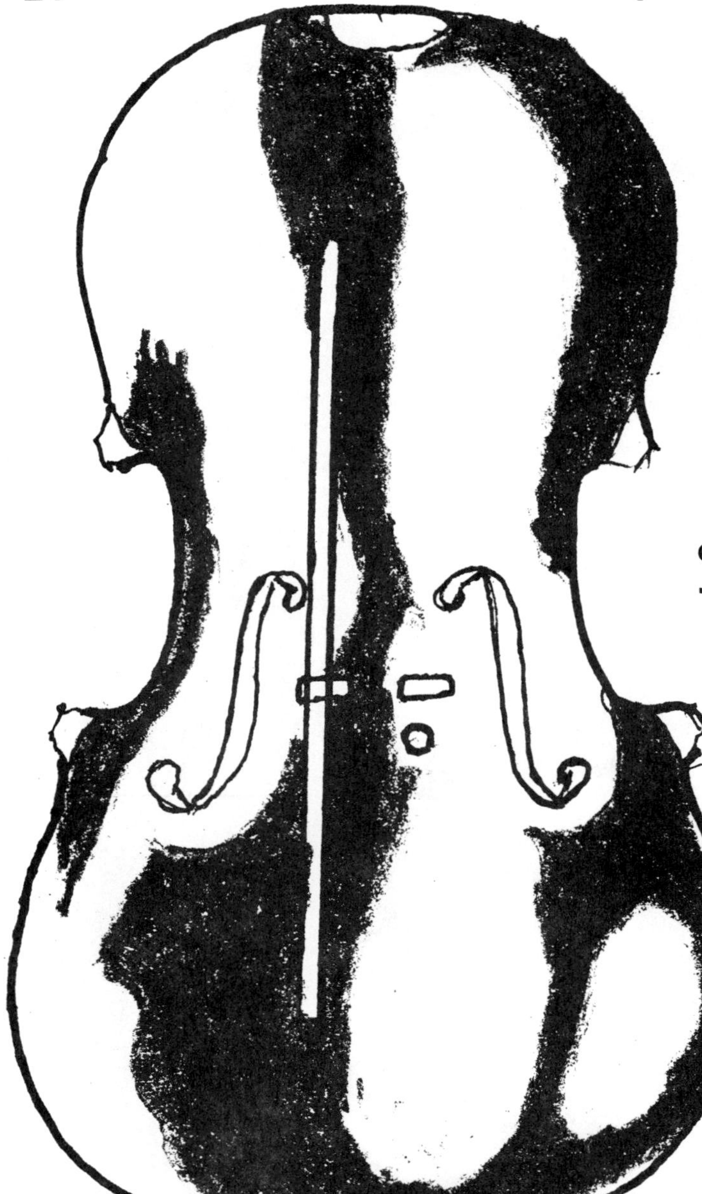

8–7

24

$F^{\#} = 740$

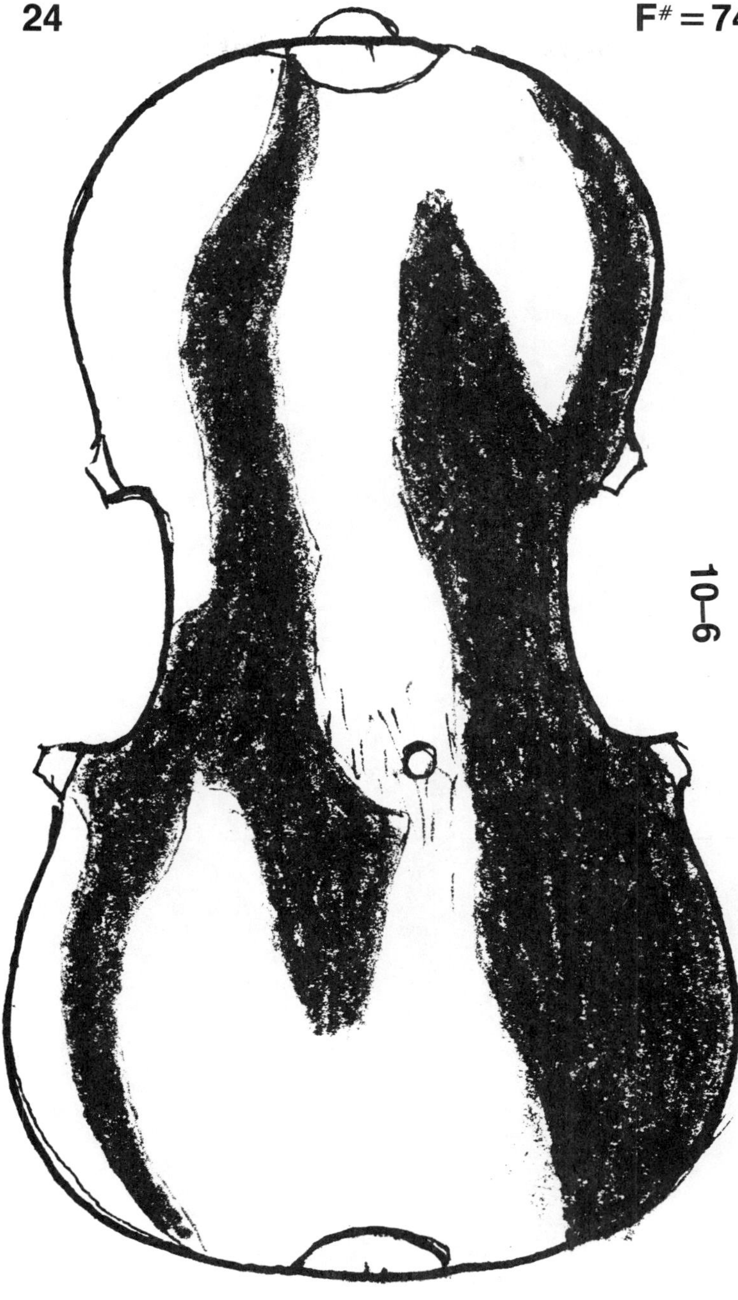

10–6

25 **G = 783.99**

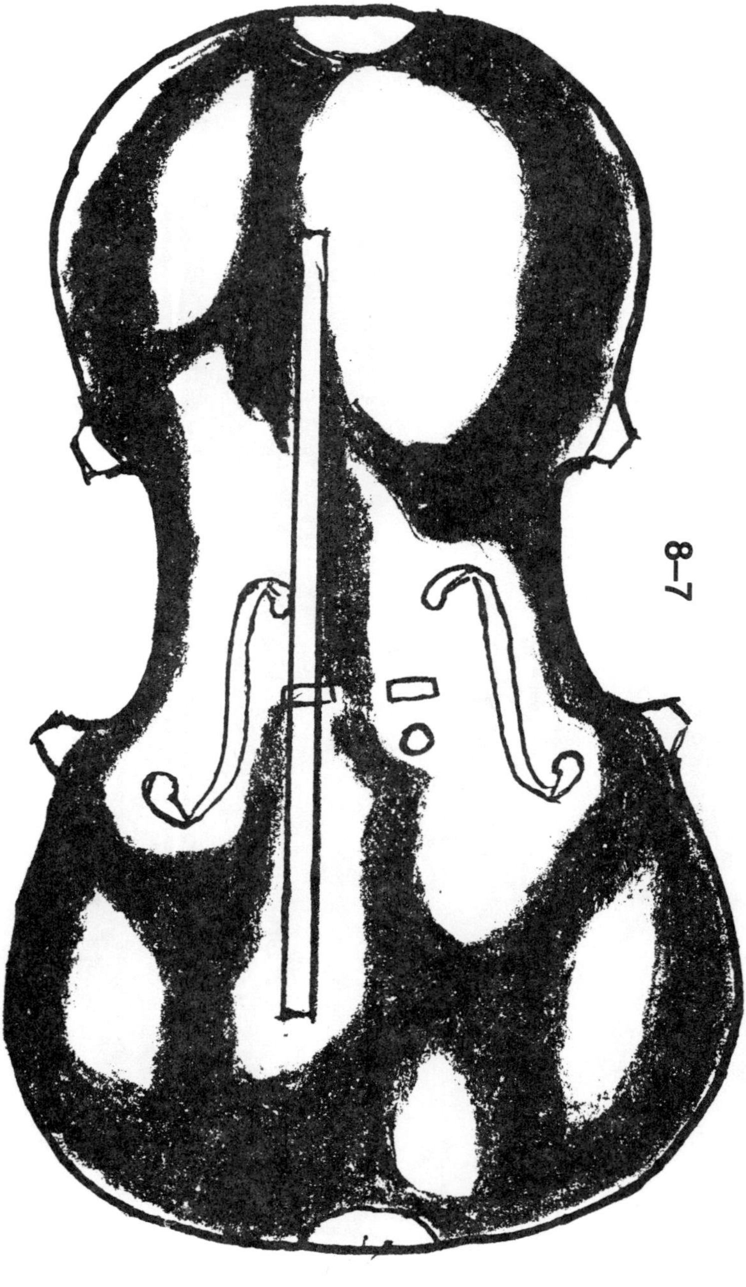

8–7

25

G = 783.99

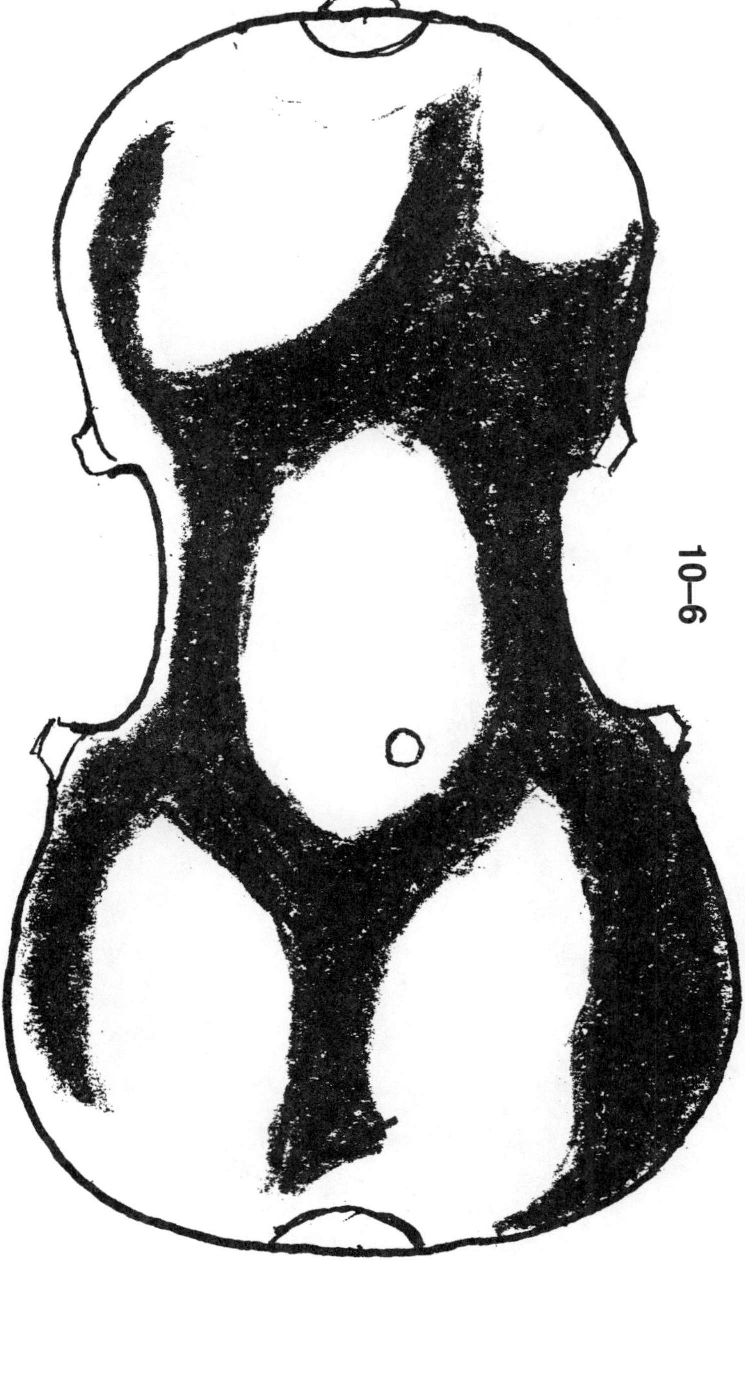

$G^{\#} = 830.61$

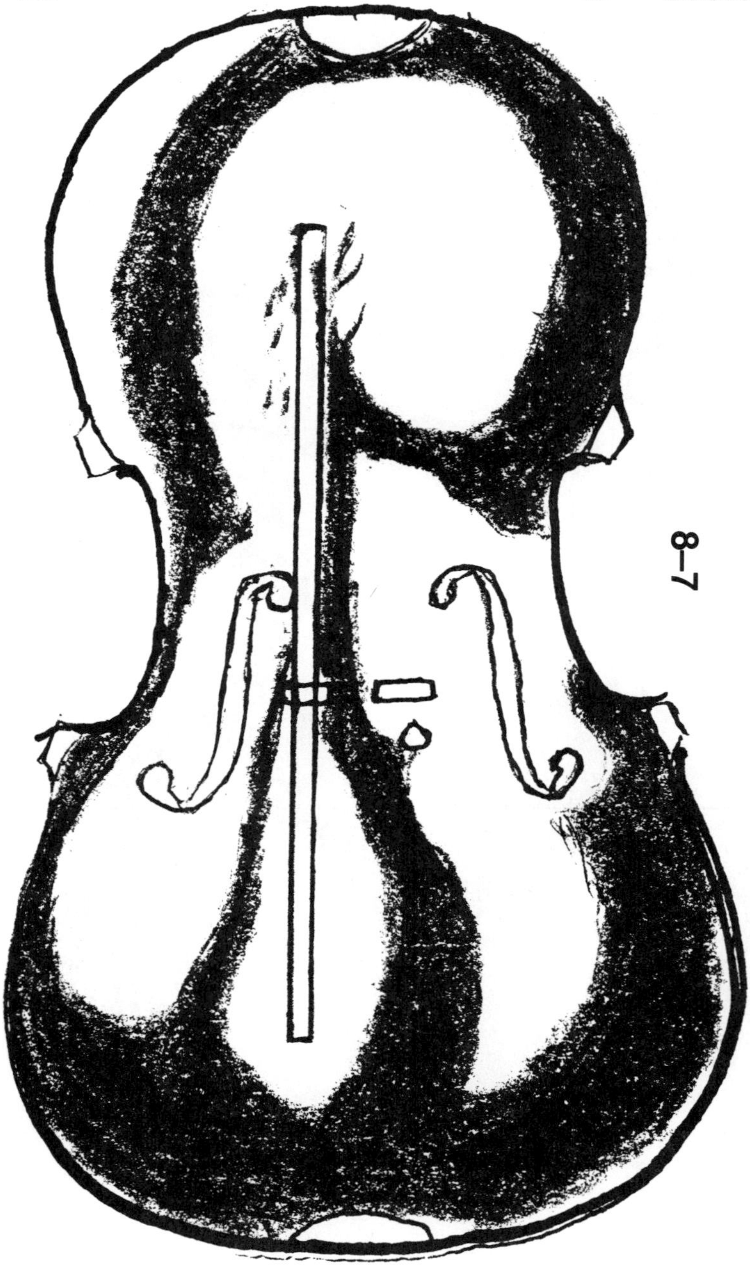

8–7

26

G# = 830.61

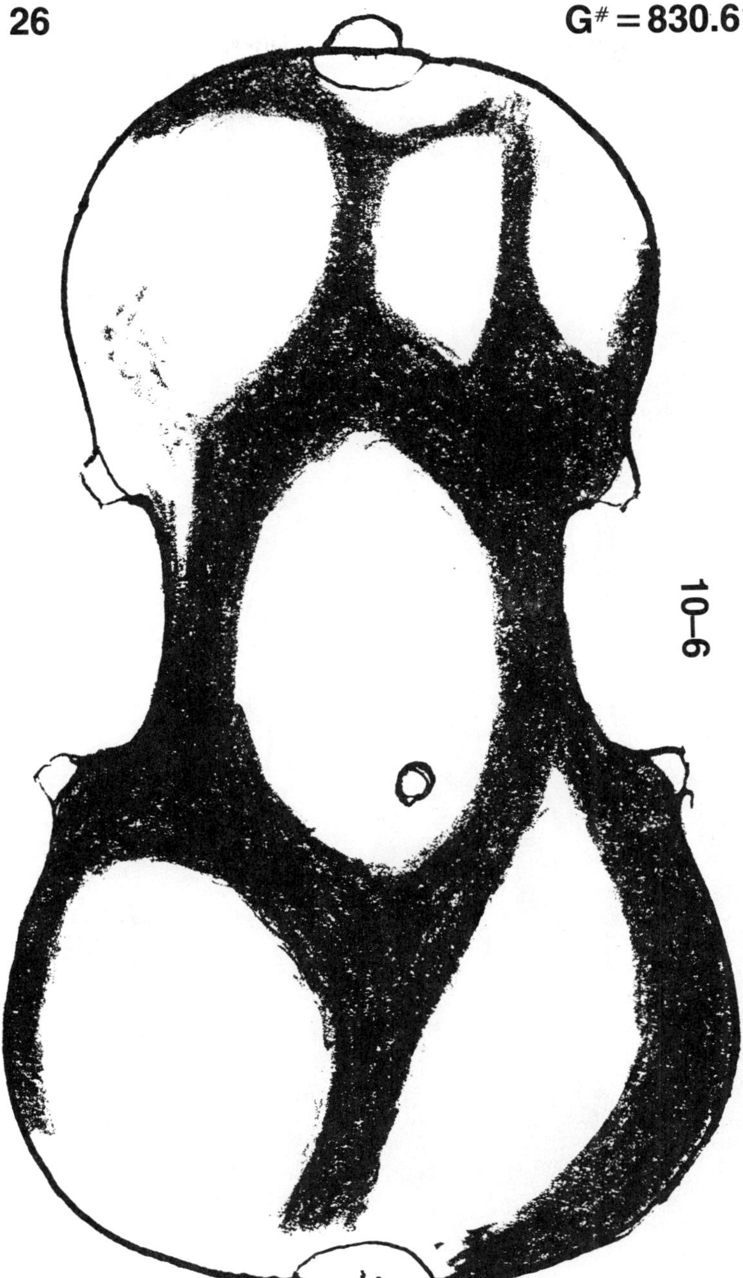

10–6

27

A = 880

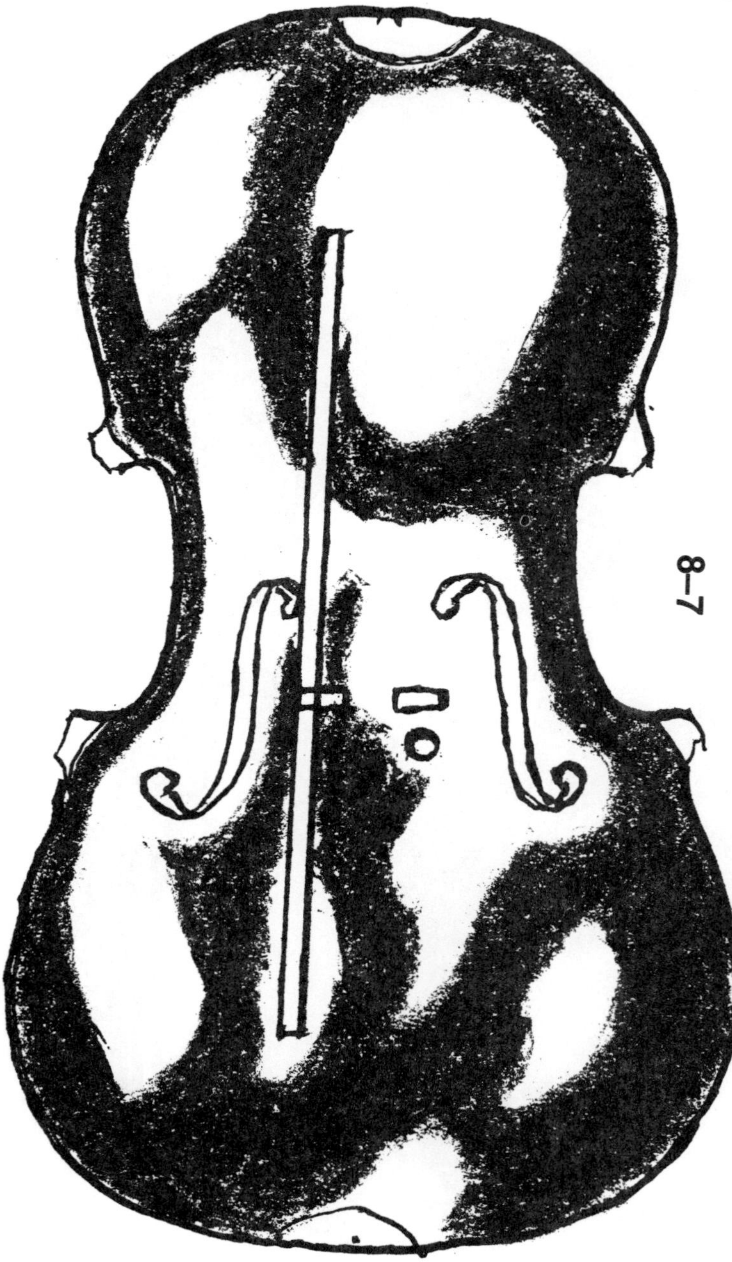

8–7

27

A = 880

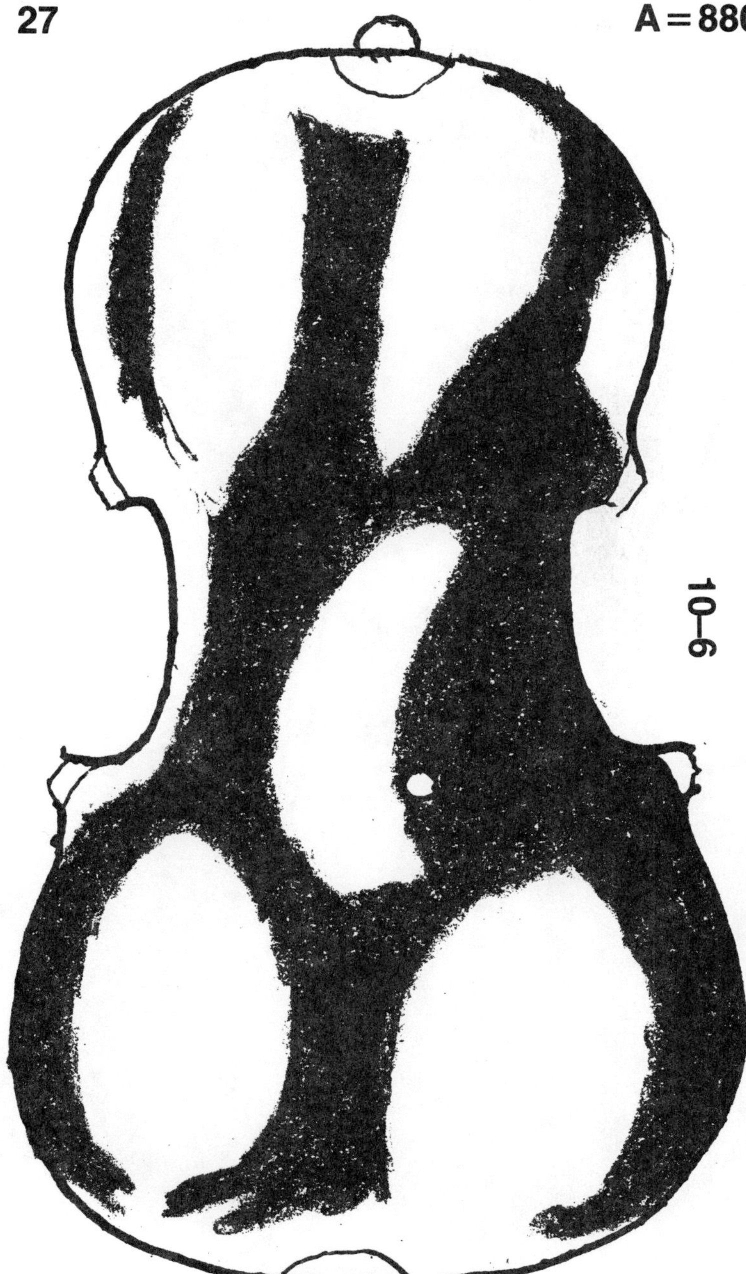

28

$A^{\#} = 932.33$

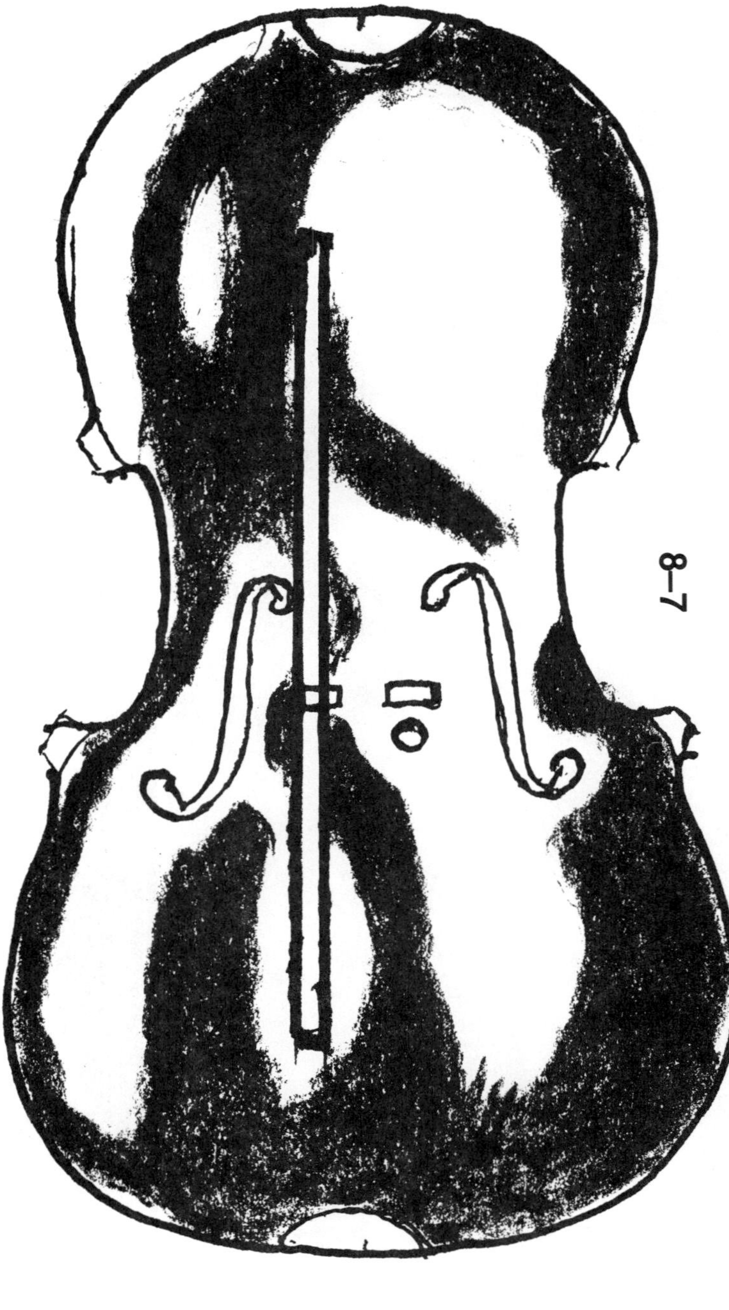

28

$A^{\#} = 932.33$

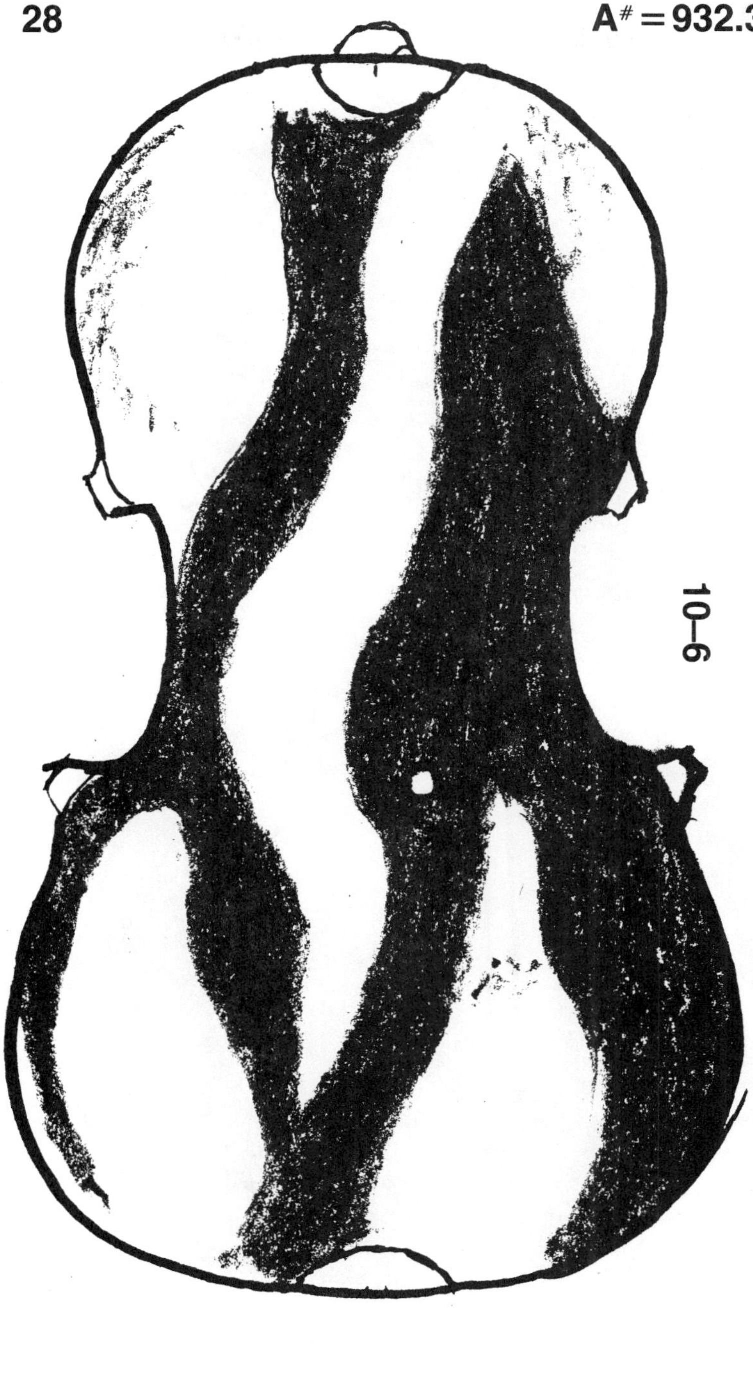

10–6

29

B = 987.77

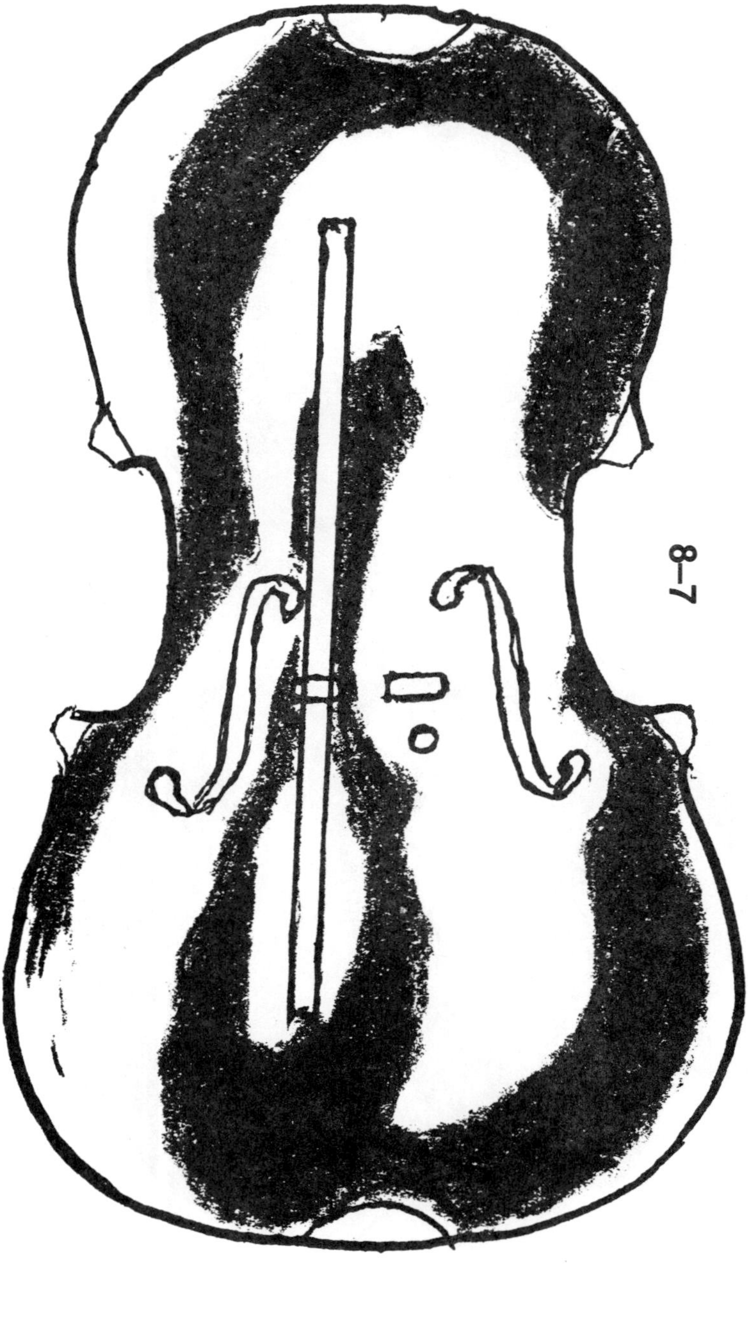

8–7

29

B = 987.77

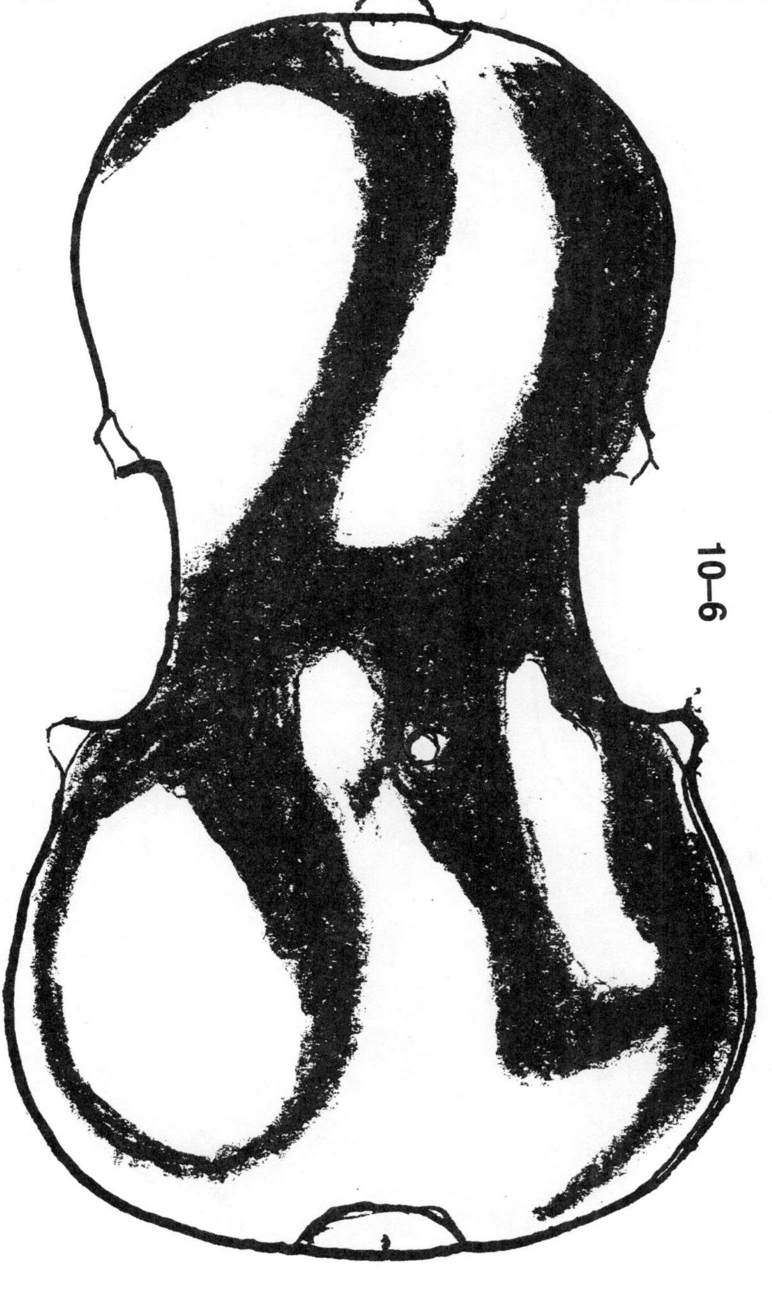

10–6

30 C = 1046.5

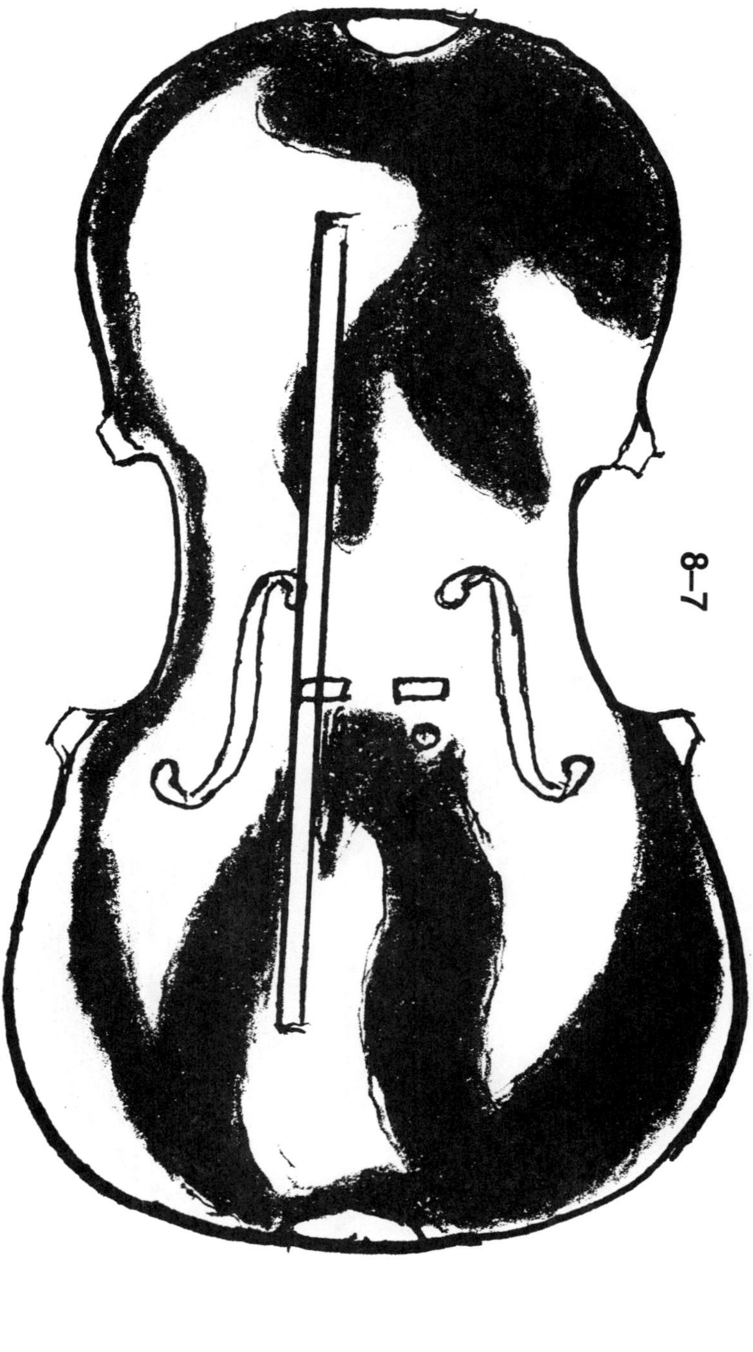

8–7

30

C = 1046.5

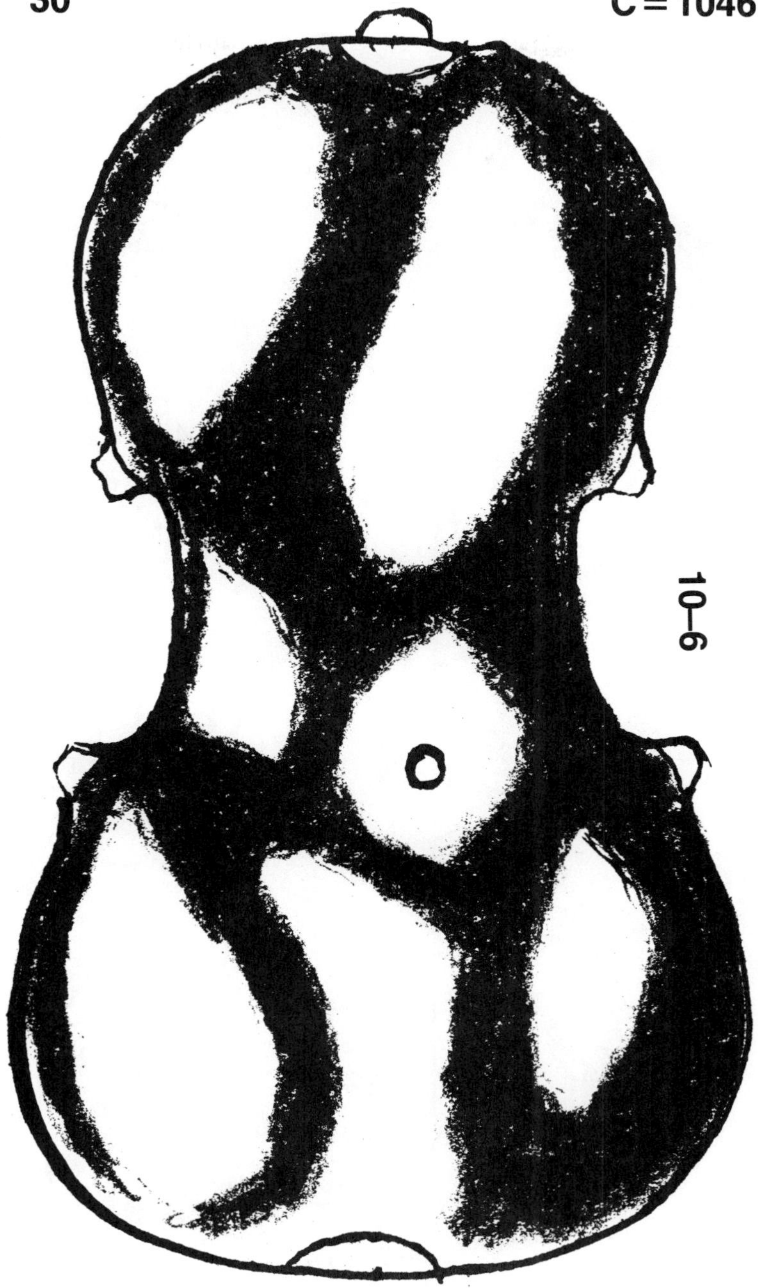

10–6

Included here are six additional sand patterns, mentioned on page 42. They were made on a plate with thicker graduation, 13/64′ at center to 7/64″ at perimeter. A violin with plates of this thickness would have a poor bass tone. G = 196, G# = 207.65, A = 220, A# = 233.08 would be very weak.

G = 196

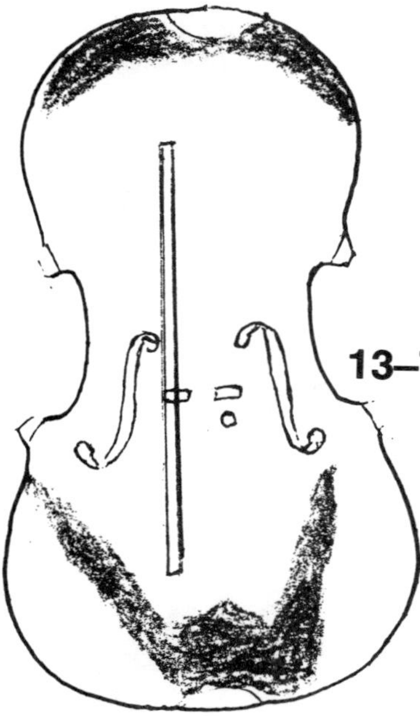

13–7

G# = 207.65

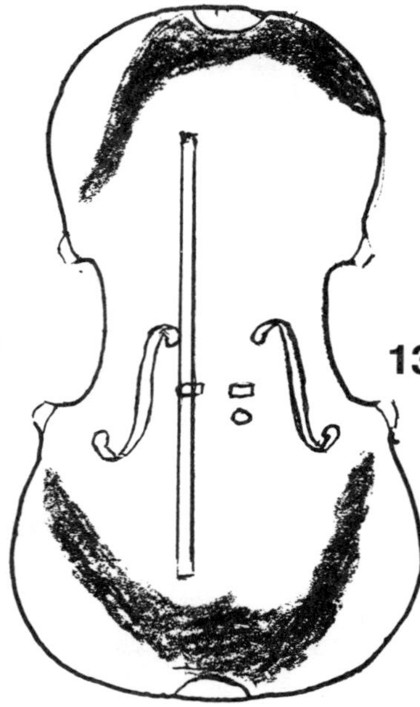

13–7

A = 220

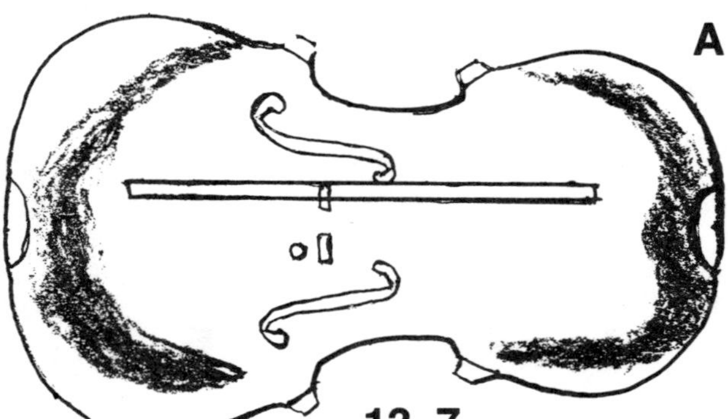

13–7

A$^{\#}$ = 233.08

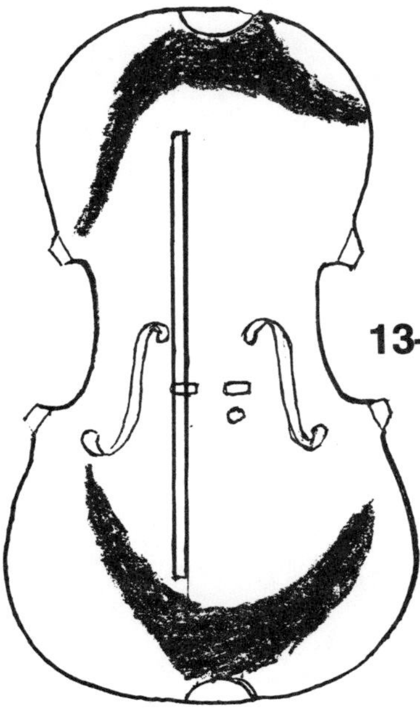

13–7

B = 246.94

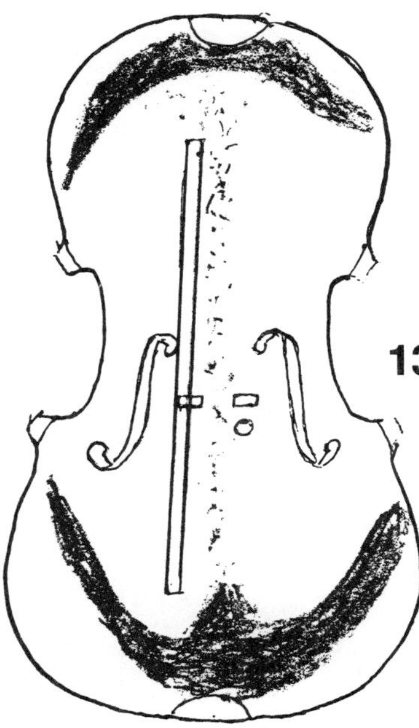

13–7

C = 261.63

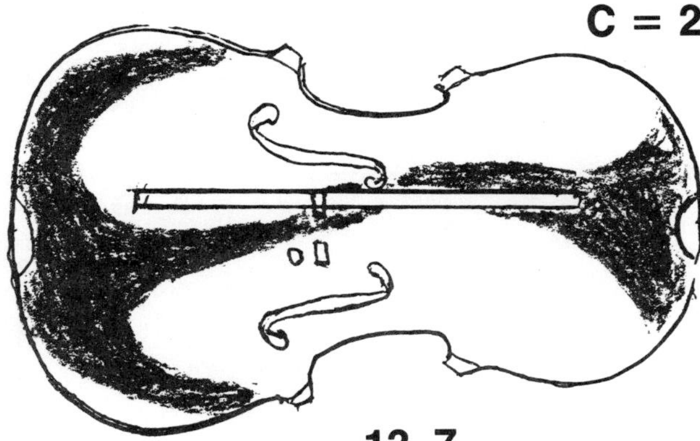

13–7

Bibliography

Culver, Chas. A. *Musical Acoustics*. New York: McGraw-Hill, 1956.

Fetis, F. J. *Anthony Stradivari, Historical and Critical Researches*. Translated by John Bishop. London: Robert Cooks & Co., 1864.

Hemholtz, Herman L. F. *Sensations of Tone, Theory of Music*. New York: Dover Publications, Inc., 1945.

Hill, Alfred E., Hill, Arthur F., Hill, W. Henry. *Antonio Stradivari, His Life and Work*. New York: Dover Publications, 1962.

Jalovek, Karel. *Italian Violin Makers*. Designed and produced by Artia for Anglo-Italian Publications. S.W. printed in Czechoslovakia. London: Anglo-Italian Publications, Ltd.

Miller, Dayton Clarence. *The Science of Musical Sounds*. New York: The MacMillan Company, 1926.

Rayleigh, John Wm. Baron. *Theory of Sound*. New York: Dover Publications, 1945.